电子科学与技术专业英语

ENGLISH FOR ELECTRONIC SCIENCE AND TECHNOLOGY

闫小兵 师建英 赵 瑞 马 蕾 编著

科 学 出 版 社

北 京

内 容 简 介

本书根据作者在电子科学与技术专业英语教学过程中长期积累的经验总结而成. 全书分为四个章节: 半导体器件、集成电路、电路, 以及学术写作. 第 1 章介绍了 pn 结、双极型器件和太阳能电池的特性; 第 2 章介绍了集成电路的发展历程、FPGA 与 ASIC、集成电路的分类以及集成电路设计工具; 第 3 章介绍了二极管和整流器电路、双极晶体管的历史和典型应用电路、信号操作处理以及直流稳压电源设计; 第 4 章介绍了科技英语文章各部分的写作方法、齐头式写作格式以及一些易错点与注意事项. 全书涵盖了微电子技术领域的基础知识, 同时又介绍了该领域的一些较新的发展状况.

本书可作为电子科学与技术专业的高年级本科生的专业英语教材, 也可供相关专业研究生以及从事相关专业领域的工程技术人员使用.

图书在版编目（CIP）数据

电子科学与技术专业英语/闫小兵等编著. —北京: 科学出版社, 2019.3
ISBN 978-7-03-053537-5

Ⅰ. ①电… Ⅱ. ①闫… Ⅲ. ①电子技术–英语–高等学校–教材
Ⅳ. ①TN

中国版本图书馆 CIP 数据核字（2017）第 117773 号

责任编辑: 赵敬伟 田轶静/责任校对: 贾娜娜
责任印制: 赵 博/封面设计: 耕者工作室

科 学 出 版 社 出版
北京东黄城根北街 16 号
邮政编码: 100717
http://www.sciencep.com

北京凌奇印刷有限责任公司印刷
科学出版社发行 各地新华书店经销
*

2019 年 3 月第 一 版　　开本: 720×1000　1/16
2024 年 5 月第五次印刷　　印张: 6 1/2
字数: 121 000
定价: 48.00 元
(如有印装质量问题, 我社负责调换)

前　　言

在信息全球化的 21 世纪,英语已经成为国际交流的重要工具. 在高等院校的理工类专业教学中,专业英语课程已然占据了极为重要的一部分,它的教学目的是衔接英语这门语言与具体理工类专业的基础知识与运用. 通过此类课程的学习,学生可以掌握科技英语技能,熟练阅读和翻译文献,撰写英文科技文章,进而了解国际上本专业的科技发展新动态. 本书的目的是满足电子科学与技术专业的本科生与研究生专业英语教学的需要.

本书主要分为四个章节:第 1 章是半导体器件方面的知识,主要包括:pn 结、双极型器件和太阳能电池的特性. 第 2 章是集成电路部分,主要包括:集成电路的发展历程、FPGA 与 ASIC 的介绍、集成电路的分类以及集成电路的设计工具. 第 3 章是电路部分的知识,主要包括二极管和整流器电路、双极晶体管的历史和典型应用电路、信号操作处理以及直流稳压电源设计. 第 4 章是写作指导部分,主要介绍科技英语文章各部分的写作方法、齐头式写作格式以及一些易错点与注意事项.

为了满足上下文的连贯性以及篇幅的限制,书中的内容做了部分删减. 为了有利于学生理解并巩固课上学习的知识,每章内容的最后都附有主要的专业词汇注释以及章节词汇与句子的翻译练习题.

本书可由教师根据自身学校的课时与专业需要选择合适的内容教学. 其他非本专业的读者也可以阅读,可提高阅读、翻译与创作科技英文文章的能力,放宽眼界.

作者在编写本书的过程中得到了多位电子科学与技术专业老师与学生的帮助,在此表示真诚的感谢. 由于编者的水平有限,疏漏之处在所难免,期待各界读者批评指正.

闫小兵
2017 年 5 月

Preface

In the 21st century of information globalization, English has become an important part of international communication. In the teaching of science and engineering in colleges and universities, the professional English course with its teaching purpose is to link the basic knowledge and application of the language and specific science and engineering majors, has already occupied an extremely important part. Through the study of such courses, students can master technical English skills, be proficient in reading and translating documents, and write English scientific and technical articles to understand the new developments in the science and technology of the international profession. The purpose of this book is to meet the needs of advanced undergraduate and postgraduate professional English teaching in electronic science and technology.

This book is divided into four chapters: Chapter 1 is the knowledge of semiconductor devices, including the pn junction, bipolar devices and solar cell characteristics. Chapter 2 is an integrated circuit part, which mainly covers the development of integrated circuits, the introduction of FPGAs and ASICs, the classification of integrated circuits, and the design tools of integrated circuits. Chapter 3 is part of the circuit, including diode and rectifier circuits, the history of dual-board transistors and typical application circuits, signal processing, and DC regulated power supply design. Chapter 4 is the writing guidance section, which mainly introduces the writing methods of each part of the scientific English article, the head-to-head writing format and some easy-to-follow points and precautions.

In order to meet the consistency of the context and the limitations of the space, the contents of the book were partially cut. Each chapter is accompanied by major professional vocabulary notes and translation exercises for chapter vocabulary and sentences, to help students understand and consolidate the knowledge of class learning.

The teaching content of this book can be selected by the teacher according to the class time and professional needs of their own school. The readers who are not in the professional might also read it, which could improve the ability of reading, translating and creating English articles in science and technology, and broaden the horizon.

In the process of writing this book, the author has received the help of many teachers and students of electronic science and technology profession. I would like to express my sincere gratitude. Due to the limited level of editors, there are inevitable omissions in the book, and readers from all walks of life are expected to criticize and correct.

Xiaobing Yan
2017.05

Contents

Preface

Chapter 1 Semiconductor Device ··········· 1
  1.1   pn Junction ··········· 1
  1.2   Bipolar Junction Transistor ··········· 6
  1.3   Solar Cells ··········· 12
  New Words and Expressions ··········· 15
  Exercises ··········· 17
  References ··········· 19

Chapter 2 Integrated Circuit (IC) ··········· 20
  2.1   The History of IC ··········· 20
  2.2   What's the IC? ··········· 22
  2.3   Classification of IC ··········· 27
  2.4   Design of ICs ··········· 31
  2.5   Low-Power Design of CMOS IC ··········· 37
  2.6   Microelectromechanical Systems ··········· 41
  2.7   Summary ··········· 47
  New Words and Expressions ··········· 47
  Exercises ··········· 49
  References ··········· 51

Chapter 3 Electric Circuit ··········· 53
  3.1   Diode and Rectifier Circuit ··········· 53
  3.2   History and Typical Application Circuit of the BJT ··········· 58
  3.3   Signal Operation and Processing ··········· 63
  3.4   Design of DC Regulated Power Supply ··········· 71
  New Words and Expressions ··········· 76
  Exercises ··········· 79

References ········· 82

Chapter 4　Writing an Academic Paper ········· 83

　4.1　Various Sections of the Academic Paper ········· 83
　4.2　Letters for Academic Communication ········· 90
　References ········· 94

Appendix ········· 95

Chapter 1 Semiconductor Device

1.1 pn Junction

The structure formed by p-type semiconductor and n-type semiconductor metallurgical contact is called pn junction. pn junction is the basic unit in almost all semiconductor devices. Actually pn junction itself is a device—rectifier, in addition to those devices that use Schottky junction. To master the physical mechanism of pn junction is the basis for learning the physics of other semiconductor devices.

1.1.1 pn Junction Space Charge Region and Contact Potential Difference

When the semiconductor materials of n-type do not contact with that of p-type, the Fermi level of the n-type materials is near the bottom of the conduction band and the Fermi level is near the top of the valence band in p-type materials, as shown in Figure 1-1(a). When the p-type materials are in contact with the n-type materials at the atomic level, and the whole system is in a state of thermal equilibrium, the Fermi level of the whole system must be equal everywhere. Otherwise, according to the modified Ohm's law, the current will flow through the system.

When the n-type and p-type semiconductor materials metallurgically contact each other, majority carriers (holes) in p-type materials diffuse to n-type side. At the same time, majority carriers (electrons) in n-type materials diffuse to p-type side. So, immovable donor ions and acceptor ions will be formed on both sides of the interface, which are described as space charges. An electric field is formed between the positive and negative space charges, which is called the built-in electric field. The built-in electric field causes the electrons and holes drift in the opposite direction of the carrier diffusion. As the diffusion goes on, the drift effect is also gradually enhanced. When the drift and diffusion effects reach the equilibrium, the movement of the carriers achieves dynamic balance so that the net current flowing through the region is zero. At this time, a stable region composed of space charges is formed at the interface between n-type and p-type semiconductor materials, which is described as space charge region (SCR), as shown in Figure 1-1(b).

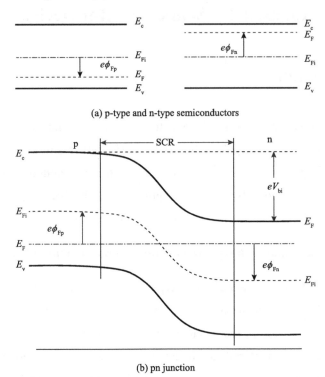

Figure 1-1 Energy band schematic diagram of p-type and n-type semiconductors and pn junction

A built-in electric field is formed in the pn junction space charge region at the equilibrium state. The electric field forms a potential difference between the n side and the p side, and the height of the built-in potential is

$$V_{bi} = |\phi_{Fn}| + |\phi_{Fp}| \tag{1-1}$$

The potential difference causes the formation of carrier diffusion barrier, so the space charge region is also called barrier region.

For pn junction that is in a state of equilibrium,

$$n_{n0} = N_D = n_i \exp\left(\frac{E_F - E_{Fi}}{kT}\right) \tag{1-2}$$

$$p_{p0} = N_A = n_i \exp\left(\frac{E_{Fi} - E_F}{kT}\right) \quad (1\text{-}3)$$

where N_D and N_A represent the effective doping concentration of n region and p region, respectively. From Figure 1-1(b), we have

$$e\phi_{Fn} = E_{Fi} - E_F = -kT \ln\left(\frac{N_D}{n_i}\right) \quad (1\text{-}4)$$

$$e\phi_{Fp} = E_{Fi} - E_F = kT \ln\left(\frac{N_A}{n_i}\right) \quad (1\text{-}5)$$

So

$$V_{bi} = |\phi_{Fn}| + |\phi_{Fp}| = \frac{kT}{q} \ln\left(\frac{N_A N_D}{n_i^2}\right) = V_T \ln\left(\frac{N_A N_D}{n_i^2}\right) \quad (1\text{-}6)$$

Contact potential difference is related to the impurity concentration, the intrinsic carrier concentration and thermal voltage.

1.1.2 Electric Field and Potential in the Depletion Layer

The built-in field is generated by the space charges in the space charge region. The relationship between the electric field intensity and the charge density is determined by the Poisson equation

$$\frac{d^2\phi(x)}{dx^2} = \frac{-\rho(x)}{\varepsilon_s} = -\frac{dE(x)}{dx} \quad (1\text{-}7)$$

where ϕ is the potential, E is the electric field strength, ρ is the charge density, and ε_s is the permittivity. According to the depletion approximation theory, from Figure 1-2(c), the charge density of the p side and the n side is

$$\rho(x) = -qN_A, \quad -x_p < x < 0 \quad (1\text{-}8a)$$

$$\rho(x) = qN_D, \quad 0 < x < x_n \quad (1\text{-}8b)$$

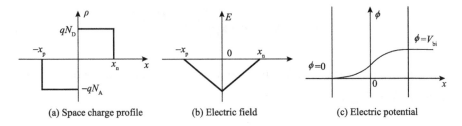

(a) Space charge profile (b) Electric field (c) Electric potential

Figure 1-2　Space charge profile, electric field and electric potential

The electric field of the p side can be obtained by integrating the $\rho(x)$ to yield

$$E = -\int \frac{qN_A}{\varepsilon_s} dx = \frac{-qN_A}{\varepsilon_s} x + C_1 = \frac{-qN_A}{\varepsilon_s}(x + x_p), \quad -x_p < x < 0 \quad (1\text{-}9a)$$

$$E = \int \frac{qN_D}{\varepsilon_s} dx = \frac{qN_D}{\varepsilon_s} x + C_2 = \frac{-qN_D}{\varepsilon_s}(x_n - x), \quad 0 < x < x_n \quad (1\text{-}9b)$$

where C_1 and C_2 are integration constants and are determined by the boundary condition $E = 0$ at $x = x_p$ and $x = x_n$, respectively. The field must be continuous at the interface ($x = 0$) between the n side and the p side, i.e.

$$E = \frac{-qN_A x_p}{\varepsilon_s} = \frac{-qN_D x_n}{\varepsilon_s} \quad (1\text{-}10)$$

The following result can be derived from Eq. (1-10)

$$N_A x_p = N_D x_n \quad (1\text{-}11)$$

Eq. (1-11) implies that the overall space charge region is electrically neutral and that the number of the positive space charge is equal to the number of negative space charge. The following expression can be derived from Eq. (1-11).

$$\frac{x_p}{x_n} = \frac{N_D}{N_A} \quad (1\text{-}12)$$

where x_n and x_p are the widths of the n-type depletion layers and the p-type depletion layers respectively. They are inversely proportional to the dopant concentration of the corresponding side; the more heavily doped side holds a smaller portion of the depletion layer.

According to the relationship between the electric field and the electric potential, integrating Eq. (1-9a) with making the voltage zero at $x = x_p$ yields as the reference point for $V = 0$ yields

$$\phi(x) = -\int E(x) dx = \int \frac{qN_A}{\varepsilon_s}(x + x_p) dx$$
$$\Rightarrow \phi(x) = \frac{qN_A}{2\varepsilon_s}(x + x_p)^2 \quad (-x_p \leqslant x \leqslant 0)$$
(1-13)

Similarly, on the n side, we integrate Eq. (1-9b) once more to obtain

$$\phi(x) = -\int E(x) dx = \int \frac{qN_D}{\varepsilon_s}(x_n - x) dx$$
$$\Rightarrow \phi(x) = \frac{qN_D}{\varepsilon_s}\left(x_n x - \frac{x^2}{2}\right) + \frac{qN_D}{2\varepsilon_s}x_n^2 \quad (0 \leqslant x \leqslant x_n)$$
(1-14)

So, we can obtain

$$V_{bi} = |\phi(x = x_n)| = \frac{q}{2\varepsilon_s}\left(N_D x_n^2 + N_A x_p^2\right)$$
(1-15)

1.1.3 Space Charge Region Width

Using Eq. (1-11) together with Eq. (1-15), one obtains

$$x_n = \left\{\frac{2\varepsilon_s V_{bi}}{q}\left[\frac{N_A}{N_D}\right]\left[\frac{1}{N_A + N_D}\right]\right\}^{1/2}$$
(1-16a)

$$x_p = \left\{\frac{2\varepsilon_s V_{bi}}{q}\left[\frac{N_D}{N_A}\right]\left[\frac{1}{N_A + N_D}\right]\right\}^{1/2}$$
(1-16b)

Then using Eq. (1-16), we can obtain the width of the space charge region as

$$W = x_n + x_p = \left\{ \frac{2\varepsilon_s V_{bi}}{q} \left[\frac{N_A + N_D}{N_A N_D} \right] \right\}^{1/2} \qquad (1\text{-}17)$$

if $N_A \gg N_D$, as in a p$^+$n junction,

$$W \approx x_n = \left\{ \frac{2\varepsilon_s V_{bi}}{q N_D} \right\}^{1/2} \qquad (1\text{-}18a)$$

if $N_A \ll N_D$, as in a pn$^+$ junction,

$$W \approx x_p = \left\{ \frac{2\varepsilon_s V_{bi}}{q N_A} \right\}^{1/2} \qquad (1\text{-}18b)$$

1.2 Bipolar Junction Transistor

At the end of 1947, Shockley, Bardeen and Brattain invented the point contact transistor at Bell Telephone Laboratories, New Jersey, USA, which was the first solid state device in the world. The invention of the transistor has opened a new era of solid state electronics. It can be said that the invention of the transistor was one of the greatest inventions of the 20th century. In 1951, the alloy junction transistor was made of germanium displaced the point contact transistor, existing for about 20 years, which had played a significant role in the practical application of solid electronic devices. In the 1970s, with the rapid development of silicon planar technology, the planar transistor fabricated by planar technology replaced the alloy junction transistor. Planar transistor includes two types, i.e. npn and pnp (Figure 1-3). In npn and pnp transistors, electrons and holes are both involved in the current transmission, so they are called bipolar junction transistor (BJT).

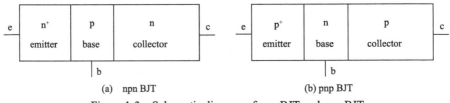

Figure 1-3   Schematic diagram of npn BJT and pnp BJT

1.2.1 Basic Structure of BJT

According to the uses, a wide variety of transistors can be divided into low frequency and high frequency transistors, low power and high power transistors, low noise transistors and high reverse bias voltage transistors and so on. According to the manufacturing process, transistors also can be divided into alloy transistor, diffusion transistor and ion injection transistor, etc.

The basic structures of various transistors are the same: p-type (or n-type) semiconductor thin layer is sandwiched between two n-type (or p-type) semiconductor layers. One of the two sides is heavily doped while the other side is lightly doped. So the transistor has two kinds of structures, i.e. npn and pnp. What's more, the thickness of the interlayer must be much smaller than the diffusion length of minority carriers.

Inside the transistor, transport process of carriers in the base region decides many physical properties, such as current gain, frequency characteristic and power characteristic. After determining geometric parameters (such as base width, emitter junction and collector junction area), the impurity distribution in the base region becomes the key factor affecting the carrier transport process. Although there are many types of transistors, for the sake of convenience, the transistor is usually divided into uniform base transistor and graded base transistor according to the base impurity distribution as one analyzes the carrier transport process in theory.

The base impurity is evenly distributed in uniform base transistors, such as the above mentioned alloy transistor. In this type of transistors, the carrier transport in the base mainly depends on the diffusion mechanism, so it is called diffusion transistor.

The impurity distribution profile is graded in graded base transistor, such as diffusion transistor. A built-in electric field caused by graded distribution of impurities is formed in the base. Due to the existence of built-in electric field, carriers not only carry out the diffusion movement, but also carry out the drift motion. The drift motion is dominant. Therefore, the graded base transistor is also called drift transistor.

1.2.2 Energy Band and Carrier Distribution of BJT

1.2.2.1 Equilibrium BJT

The energy band and carrier distribution of the uniform base transistor without applied voltage (i.e., the equilibrium state) is shown in Figure 1-4. The emitter, base and collector are evenly distributed. The doping concentration of the emitter is the highest, the doping concentration of the base is lower than that of emitter, and that of the collector is the lowest. U_{De} and U_{Dc} denote the contact potential difference of the emitter and collector junction. At the equilibrium state, the transistor has a uniform Fermi level.

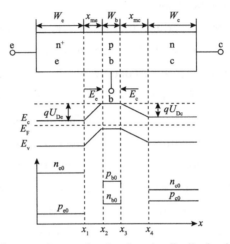

Figure 1-4 Schematic diagram of energy band and carrier distribution in an equilibrium transistor

1.2.2.2 Non-equilibrium BJT

When the transistor is operating under the amplification configuration, the emitter junction is forward biased (represented by U_e) and the collector junction is applied to reverse bias (denoted by U_c). At this time, the transistor emitter and collector junction are in the non-equilibrium condition. Therefore, The Fermi energy level is not consistent throughout the system. The schematic diagram of energy band is shown in Figure 1-5 (b).

Under the amplification conditions, the distribution of carriers in BJT is shown in Figure 1-5(c). A forward bias of U_e reduces the barrier height from qU_{De} to $q(U_{De}-U_e)$. This decreases the drift field and breaks the balance between diffusion and drift that exists at zero bias. Therefore, electrons can diffuse from the emitter (the n side) into the base (the p side). This process is called minority carrier injection. Electrons (minority) accumulate at the base boundary and diffuse into the base region. Simultaneously, the

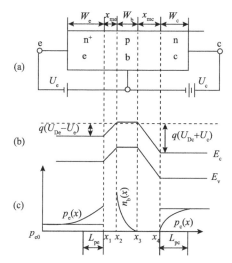

Figure 1-5 Energy band and carriers distribution under non-equilibrium conditions

recombination of electrons and holes occurs, and finally a stable distribution is formed, denoted by $n_b(x)$. Also, holes (minority) of base region are injected into the emitter region, and finally a stable distribution is formed, denoted by $p_e(x)$. For the collector junction, the carrier concentration at collector junction boundary will be reduced to zero due to its reverse bias extraction. The minority carrier concentration distribution in the collector region is represented by $p_c(x)$.

1.2.3 Carriers Transport and Current Formation

The main function of the transistor is to amplify the electrical signal. A pn junction has only rectifier function instead of amplifying electrical signals. However, when two pn junctions are very close to each other, they have the ability to amplify the current due to the interaction. Below, taking the npn transistor as an example, the principle of the current amplification of the transistor is analyzed.

The transistor discussed below generally refers to the uniform base transistor (unless otherwise specified), and has the following assumptions:

1) The emitter and the collector widths are far greater than the minority carrier diffusion length, but base width is far less than the minority carrier diffusion length.

2) The resistivity of the emitter and the collector is low enough and the applied voltage is all falling in the barrier region. There is no electric field outside the barrier region.

3) The space charge region widths of emitter junction and collector junction are much smaller than the diffusion length. There is no generation and recombination there.

4) Impurities in three areas are uniform distribution. Surface effects are not considered. The carriers transport only in one dimensional direction.

5) Low level injection, that is, the non-equilibrium minority carrier concentration injected into base is far less than the majority carrier concentration.

6) The emitter junction and collector junction are ideal abrupt junctions and their areas are equal.

1.2.3.1　Carriers transport

According to Figure 1-6(b), the carrier transport process can be divided into three parts.

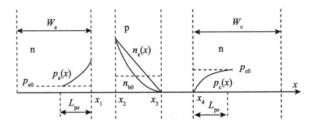

(a) Concentration distribution of minority carrier

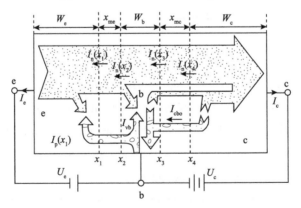

(b) Diagram of carriers transport in BJT

Figure 1-6　Concentration distribution of minority carrier and diagram of carriers transport in BJT

(1) Emitter junction is forward biased emitting electrons

Since the emitter junction is forward biased, the applied electric field favors the diffusion movement of the majority carriers, and therefore the majority carriers (electrons) in the highly doped emitter will diffuse into the base region (inject). The majority carriers (holes) in the base also diffuse to the emitter and recombine with some of the electrons in the emitter.

(2) Transport and recombination of carriers in base

A small fraction of electrons injected into the base region are recombined with the majority carriers (holes) in p-type base region. However, since the hole density of the low-doped base is low and the thickness of base is very thin, few electrons injected into base disappear by recombination, and most of the electrons can transport to the boundary between the collector junction and the base.

(3) Collector junction is reverse biased-collecting electrons

Because the collector junction is reverse biased, it will sweep the electrons to the collector, which are injected from the emitter junction and diffuse to the collector junction boundary. Then, electrons can be collected by the collector.

In addition, since the collector junction is reverse biased, minority carriers (holes) in the collector and minority carriers (electrons) in the base are swept toward the base and the collector, respectively.

1.2.3.2 BJT terminal current

The BJT terminal current includes I_c, I_e and I_b.

(1) Emitter terminal current I_e

Emitter terminal current I_e is composed of two internal currents: one part is electron diffusion current $I_n(x_1)$ arised from a forward biased emitter junction that is injected from emitter to base. Most of these currents can transport the collector junction and become the major part of collector terminal current I_c. The other part is the hole diffusion current $I_p(x_1)$ ascribed to the forward biased emitter junction that is injected from base to emitter. The current $I_p(x_1)$ does not contribute to the collector terminal current I_c but it is a part of the base terminal current I_b. Therefore,

$$I_e = I_p(x_1) + I_n(x_1) \tag{1-19}$$

(2) Base terminal current I_b

Base terminal current I_b is composed of three parts of current: one part is the recombination current I_{vb} occurred in base, which represents the recombination current of electrons injected into base and holes existed in base; the second part is the hole diffusion current $I_p(x_1)$ injected from the base into the emitter; the third part is reverse saturation current I_{cbo} across the reverse biased collector junction. So

$$I_b = I_p(x_1) + I_{vb} - I_{cbo} \tag{1-20}$$

(3) Collector terminal current I_c

Collector terminal current is composed of two parts of current: the dominant part is the electron diffusion current $I_n(x_3)$, which equals the current injected into base minus the base recombination current. A tiny fraction of I_c is the reverse saturation current I_{cbo}. So the collector terminal current I_c is given by

$$I_c = I_n(x_4) + I_{cbo} \tag{1-21}$$

1.3 Solar Cells

In 1947, Shockley and other two scientists working at Bell Telephone Laboratories in USA invented the transistor and inaugurated a new era in semiconductor science and technology field. In 1954, Pearson et al. who came from the same laboratory successfully developed the first silicon (Si) single crystal pn junction solar cells in the world, which ushered in a new epoch in modern photovoltaic technology development. The invention of the transistor originated from the solid energy band theory based on the quantum mechanics. The successful development of the solar cell benefited from the pn junction principle of the semiconductor physics. The invention of the transistor made outstanding contributions for the development of contemporary semiconductor and microelectronics technology. The birth of solar cells made a contribution for the development of modern new energy technology.

So far, the development of modern photovoltaic technology has gone through 60 years of glorious history. In order to have a clearer understanding for the scientific connotation of various types of solar photovoltaic principles, this chapter first makes a brief review of the development history of solar cells.

1.3.1 Inorganic Solid pn Junction Solar Cells

What is the inorganic solid pn junction solar cell? In short, it is photovoltaic device that is made of single crystal or polycrystalline semiconductor materials and uses pn junction as the light absorption active region. Such solar cells include single crystal Si solar cells, polycrystalline Si solar cells and compound GaAs solar cells. It is well known that Si and GaAs are the preferred materials for the fabrication of various microelectronic devices and ICs. In addition, due to the forbidden band width of Si and GaAs materials is in the best energy absorption range of the solar spectrum, they are also excellent photovoltaic materials. In particular, monocrystalline Si and polycrystalline Si solar cells play a dominant role in today's development of photovoltaic industry, due to their high energy conversion efficiency and mature production process. Figure 1-7 shows the structure and energy band of single Si pn junction solar cell. After the solar cells absorb light energy, the photogenerated carriers diffuse into the junction area and are separated by built-in electric field, finally are collected by the electrode.

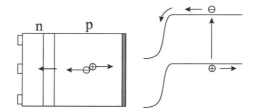

Figure 1-7 Structure and energy band of single Si pn junction solar cell

1.3.2 Thin-Film Solar Cells

The purpose of developing thin-film solar cells is to reduce the production cost of solar cells. If thin-film solar cells with a layer thickness of only a few microns can be fabricated on inexpensive substrates such as glass, stainless steel or plastic, their prices will be significantly reduced. However, the current energy conversion efficiency of various types of thin-film solar cells is not high enough, so to improve the conversion efficiency as much as possible under the premise of reducing the production cost is another major goal. High efficiency and low cost are always the theme of development of modern photovoltaic devices. Thin-film solar cells mainly include Si-based thin-film solar cells, Cu(In,Ga)Se$_2$ (CIGS) thin-film solar cells and CdTe thin-film solar cells. Figure 1-8(a), (b) and (c) show the structures and energy band diagrams of the three types of thin-film solar cells, namely, amorphous silicon (a-Si:H), Cu(In,Ga)Se$_2$ and CdTe. The typical a-Si:H single-junction thin-film solar cell is a p-i-n structure, the intrinsic (i) layer with a certain width between the p-type and n-type a-Si:H thin-film acts as the light absorption active region. The strong built-in electric field formed in the intrinsic layer can effectively separate the electrons and holes couple generated by the light, thereby improving the conversion efficiency of the solar cell. For the latter two thin-film solar cells, the window layer is made of wide-band-gap CdS material, which allows more incident light to be absorbed by Cu(In,Ga)Se$_2$ and CdTe active regions.

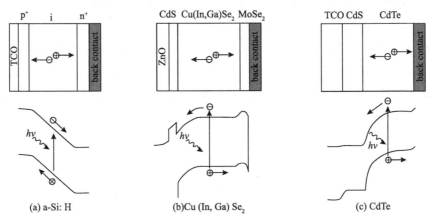

Figure 1-8 Structure and energy band diagram of a-Si: H, Cu (In, Ga) Se$_2$ and CdTe thin-film solar cell

1.3.3 New Concept Solar Cells with High Efficiency

In order to break through the current bottleneck in the development of photovoltaic technology, the new concept solar cells with ultra-high conversion efficiency are proposed. In short, this is a kind of high efficiency, low cost, long life, high reliability and non-toxic third-generation solar cells, which are nearly ideal "environmental friendly, green and efficient" modern photovoltaic devices. These new concept solar cells mainly include multi-junction stacked solar cells, quantum well solar cells, nanostructured solar cells, quantum dot intermediate band solar cells, quantum dot multi-exciton solar cells, hot-carrier solar cells and surface plasmon enhanced solar cells.

New Words and Expressions

pn junction	pn 结
Schottky junction	肖特基结
metallurgy	冶金
semiconductor device	半导体器件
rectifier	整流
n-type	n 型
p-type	p 型
Fermi level	费米能级
conduction band	导带
valance band	价带
thermal equilibrium	热平衡
non-equilibrium	非平衡
Ohm's law	欧姆定律
majority carrier	多数载流子
minority carrier	少数载流子
donor ions	施主离子
acceptor ions	受主离子
built-in electric field	内建电场
drift	漂移
diffusion	扩散
space charge region	空间电荷区
contact potential difference	接触电势差
barrier	势垒
intrinsic carrier concentration	本征载流子浓度
impurity concentration	杂质浓度
permittivity	介电常数
Poisson equation	泊松方程
depletion layer	耗尽层
boundary condition	边界条件

bipolar junction transistor	双极结型晶体管
manufacturing process	制造工艺
transport	输运
current gain	电流增益
impurity distribution	杂质分布
uniform base transistor	均匀基区晶体管
graded base transistor	缓变基区晶体管
emitter	发射区
base	基区
collector	集电区
forward biased	正偏
reverse bias	反偏
recombination	复合
current	电流
resistivity	电阻率
inject	注入
terminal current	端电流
solar cell	太阳电池
inorganic solid	无机固态
single crystal	单晶
polycrystalline	多晶
monocrystalline	单晶
plastic	塑料
substrate	衬底
glass	玻璃
stainless steel	不锈钢
thin-film solar cell	薄膜太阳电池
photovoltaic device	光伏器件
amorphous silicon	非晶硅
intrinsic layer	本征层
new concept solar cell	新概念太阳电池
long life	长寿命

high reliability	高稳定性
non-toxic	无毒
environmental friendly	环境友好
quantum well solar cell	量子阱太阳电池
quantum dot intermediate band solar cell	量子点中间带太阳电池
quantum dot multi-exciton solar cell	量子点多激子太阳电池
hot-carrier solar cell	热载流子太阳电池
surface plasmon enhanced solar cell	表面等离子增强太阳电池

Exercises

1. 请将下述词语翻译成英文

pn 结	肖特基结	多数载流子	少数载流子
施主离子	受主离子	空间电荷区	扩散
漂移	复合	接触电势差	内建电场
正偏	反偏	耗尽层	晶体管
平衡	非平衡	载流子输运	端电流
薄膜太阳电池	新概念太阳电池	中间带太阳电池	太阳电池
多激子太阳电池	光生载流子		

2. 请将下述词语翻译成中文

semiconductor device	rectifier	thermal equilibrium
Fermi level	built-in electric field	drift
impurity concentration	permittivity	solar cell
contact potential difference	stainless steel	amorphous silicon
environmental friendly		

3. 请将下述短文翻译成中文

1) The structure formed by p-type semiconductor and n-type semiconductor metallurgy contact is called pn junction. pn junction is the basic unit in almost all semiconductor devices.

2) When the n-type and p-type semiconductor materials metallurgically contact each other, majority carriers (holes) in p-type materials diffuse to n-type side. At the

same time, majority carriers (electrons) in n-type materials diffuse to p-type side. So, immovable donor ions and acceptor ions will be formed on both sides of the interface, which are described as space charges.

3) At the end of 1947, Shockley, Bardeen and Brattain invented the point contact transistor at Bell Telephone Laboratories, New Jersey, USA, which was the first solid state device in the world.

4) In the 1970s, with the rapid development of silicon planar technology, the planar transistor fabricated by planar technology had replaced the alloy junction transistor. Planar transistor includes two types, i.e. npn and pnp.

5) The basic structures of various transistors are the same: p-type (or n-type). Semiconductor thin layer is sandwiched between two n-type (or p-type) semiconductor layers. One of two sides is heavily doped while the other side is lightly doped.

6) Although there are many types of transistors, for the sake of convenience, the transistor is usually divided into uniform base transistor and graded base transistor according to the base impurity distribution as one analyzes the carriers transport process in theory.

7) The base impurity is evenly distributed in uniform base transistors, such as the above mentioned alloy transistor. In this type of transistors, the carrier transport in the base mainly depends on the diffusion mechanism, so it is called diffusion transistor.

8) The impurity distribution profile is graded in graded base transistor, such as diffusion transistor. A built-in electric field caused by graded distribution of impurities is formed in the base. Due to the existence of built-in electric field, carriers not only carry out the diffusion movement, but also carry out the drift motion. The drift motion is dominant.

9) The main function of the transistor is to amplify the electrical signal. A pn junction has only rectifier function instead of amplifying electrical signals. However, when two pn junctions are very close to each other, they have the ability to amplify the current due to the interaction.

10) The resistivity of the emitter and the collector is low enough and the applied voltage is all falling in the barrier region. There is no electric field outside the barrier region.

11) The emitter and the collector widths are far greater than the minority carrier

diffusion length, but base width is far less than the minority carrier diffusion length.

12) Low level injection, that is, the non-equilibrium minority carrier concentration injected into base is far less than the majority carrier concentration.

13) In 1954, Pearson et al. who came from the same laboratory successfully developed the first silicon (Si) single crystal pn junction solar cells in the world, which ushered in a new epoch in modern photovoltaic technology development.

14) So far, the development of modern photovoltaic technology has gone through 60 years of glorious history.

15) What is the inorganic solid pn junction solar cell? In short, it is a photovoltaic devices that are made of single crystal or polycrystalline semiconductor materials and use pn junction as the light absorption active region.

16) The purpose of developing thin-film solar cells is to reduce the production cost of solar cells. If thin-film solar cells with a layer thickness of only a few microns can be fabricated on inexpensive substrates such as glass, stainless steel or plastic, their prices will be significantly reduced.

17) In short, this is a kind of high efficiency, low cost, long life, high reliability and non-toxic third-generation solar cells, which are nearly ideal "environmental friendly, green and efficient" modern photovoltaic devices.

References

陈星弼, 张庆中, 陈勇编. 2011. 微电子器件. 3 版. 北京: 电子工业出版社.
彭英才, 赵新为, 李晓伟. 2015. 现代光伏器件物理. 北京: 科学出版社.
施敏, 伍国钰. 半导体器件物理. 2008. 耿莉, 张瑞智, 译. 西安: 西安交通大学出版社.
Anderson B L, Anderson R L. 2008. 半导体器件基础. 邓宁, 田立林, 任敏, 译. 北京: 清华大学出版社.
Pierret R F. 2010. 半导体器件基础. 黄如, 王漪, 王金延, 等译. 北京: 电子工业出版社.

Chapter 2   Integrated Circuit (IC)

2.1   The History of IC

Our world is full of ICs. You find several of them in computers, telephones, cameras and so on. For example, most people have probably heard about the microprocessor. The microprocessor is an IC that processes all information in the computer. It keeps track of what keys are pressed and if the mouse has been moved. It counts numbers and runs programs, games and the operating system. ICs are also found in almost every modern electrical device such as cars, television sets, CD players, cellular phones, etc. But what is the history of the IC?

In 1821, Thomas Seebeck discovered semiconductor properties of PbS.

In 1833, Michael Faraday reported on conductivity temperature dependence for a new class of materials—semiconductors.

In 1873, W. Smith discovered light sensitivity of semiconductors.

In 1874, Ferdinand Braun, a German scientist, discovered that crystals could conduct current in one direction under certain conditions. This phenomenon is called rectification.

In 1875, Werner von Siemens invented a selenium photometer.

In 1878, Alexander Graham Bell used this device for a wireless communication system.

In 1895, the Italian Guglielmo Marconi first showed a new technology invented by Nikola Tesla through radio signals. This was the beginning of wireless communication. Crystal detectors were used in radio receivers. They are able to separate the carrier wave from the part of the signal carrying the information.

In 1904, John Ambrose Fleming, an English physicist, devised the first practical electron tube known as the "Fleming Valve". Vacuum tube was discovered in 1904. The first digital computer ENIAC (Figure 2-1), for example, was a huge monster that weighed over thirty tons, and consumed 200kW of electrical power. It had around 18,000 vacuum tubes that constantly burned out, making it very unreliable.

Figure 2-1   The first digital computer ENIAC

In 1906, Lee de Forest, an American scientist, added the third electrode (called a grid) to the electron tube, which is now called a triode. This is a network of small wires around the vacuum tube cathode. Thus, the amplifying vacuum tube, the most recent ancestor of the transistor, was born.

In 1907, Round demonstrated the first LED (using SiC).

In 1938, Schottky proposed M-S barrier theory (model).

In 1947, Bardeen, Brattain, and Shockley discovered a Bipolar Junction transistor and the Modern Age began. It was considered a revolution. Being small, fast, reliable and effective, it quickly replaced the vacuum tube.

In 1949, Shockley proposed pn junction and BJT theory.

In 1958, at Texas Instruments, Jack Kilby was probably most famous for his invention of the IC (as shown in Figure 2-2), for which he received the Nobel Prize in Physics in the year 2000.

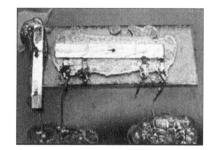

Figure 2-2   The first IC

In 1961, the first commercially available IC came from the Fairchild Semiconductor Corporation. All computers then started to be made of chips instead of the individual transistors and their accompanying parts.

In 1963, Wanlass and Sah introduced CMOS technology.

In 1965, Moore's law, proposed by Gordon Moore (Figure 2-3), co-founder of Intel, who predicted that every 65,000 electronic components would be integrated in one single chip. Moore's law says that the number of transistors on a chip doubles every 18 months or so.

The IC has come a long way since Jack Kilby's first prototype. His idea founded a new industry and is the key element behind our computerized society. Today the most advanced circuits contain several hundred millions of components on an area no larger than a fingernail. The transistors on these chips are around 90nm, which means that you could fit hundreds of these transistors inside a red blood cell.

Figure 2-3 Gordon Moore

Each year computer chips become more powerful yet cheaper than the year before. Gordon Moore, one of the early IC pioneers and founders of Intel once said, "If the auto industry advanced as rapidly as the semiconductor industry, a Rolls Royce would get a half a million miles per gallon, and it would be cheaper to throw it away than to park it."

2.2 What's the IC?

An electric circuit is made from different electrical components such as transistors, resistors, capacitors and diodes that are connected to each other in different ways. These components have different behaviors.

The transistor acts like a switch. It can turn electricity on or off, or it can amplify current. It is used for example in computers to store information, or in stereo amplifiers to make the sound signal stronger. The resistor limits the flow of electricity and gives us the possibility to control the amount of current that is allowed to pass. The capacitor collects electricity and releases it all in one quick burst, just like an instance of cameras where a tiny battery can provide enough energy to fire the flashbulb. The diode stops electricity under some conditions and allows it to pass only when these conditions change. These components are like the building blocks in an electrical construction kit. It depends on how the components are put together when building the circuit. Everything from a burglar alarm to a computer microprocessor can be constructed.

When building a circuit, it is very important that all connections are intact. Before the IC, assembly workers had to construct circuits by hand, soldering each component in place and connecting them with metal wires. Engineers soon realized that manually assembling the vast number of tiny components needed in, for example, a computer would be impossible, especially without generating a single faulty connection. Another problem was the size of the circuits. If the components of the computer were too large or the wires interconnecting them too long, the electric signals couldn't travel fast enough through the circuit, thus making the computer too slow to be effective.

IC uses the semiconductor production process, with many transistors and resistors, capacitors and other components in a small silicon wafer. The components are combined into a complete electronic circuit with the wiring method of multilayer. Figure 2-4 are some samples of ICs.

Figure 2-4   Samples of ICs

IC is also called microcircuit, microchip and chip. ICs have the advantages of small volume, light weight, less leading wire and welding points, long service life, high reliability, good performance, low cost, convenient for mass production.

The resulting small chip of semiconductor materials on which a transistor or diode is fabricated can be encased in a small plastic package for protection against damage and contamination from the outside world. Small wires are connected within this package between the semiconductor sandwich and pins that protrude from the package to make electrical contact with other parts of the IC. Once you have several discrete transistors, digital logic can be built by directly wiring these components together. The circuit will function. Any substantial amount of digital logic will be very bulky because several transistors are required to implement each of the various types of logic gates.

When an IC is designed and fabricated, it generally follows one of two main transistor technologies: bipolar or metal-oxide semiconductor (MOS). Bipolar processes create BJTs, whereas MOS processes create FETs. Bipolar logic was more common before the 1980s, but MOS technologies have since accounted the great majority of digital logic ICs. n-channel FETs are fabricated in a NMOS process, and p-channel FETs are fabricated in a PMOS process. In the 1980s, complementary-MOS, or CMOS, became the dominant process technology and remains so to this day. CMOS IC incorporated both NMOS and PMOS transistors.

2.2.1 ASIC

ASIC is an abbreviation for application-specific IC. It is a kind of IC which designed specifically for a particular application or purpose. Compared with programmable logic devices or standard logic ICs, ASIC can improve the circuit's speed because it is specifically designed only to do one thing and it does very well. It also can become smaller and use less electricity. The disadvantage of ASIC is that it is more expensive to be designed and manufactured, particularly if only a few units are needed.

The term ASIC has a wide variety of associations. Strictly speaking, it simply refers to an IC that has been designed for a particular application. This defines a portion of the semiconductor market. Other market segments include memories, microprocessors, and field programmable gate arrays(FPGAs).

Generally, ASICs are designed at the register-transfer level (RTL) in Verilog or VHDL, specifying the flow of data between registers and the state to store in registers.

Commercial EDA tools are used to map the higher level and route wires. It is much easier to migrate ASIC designs to a new process technology, compared to custom design which has been optimized for a specific process at the gate or transistor-level. ASIC designers generally focus on high level designs choices, at the micro-architectural level for example.

With this broader context, let us pause to note that the use of the term ASIC can be misleading: it most often refers to an IC produced through a standard cell ASIC methodology and fabricated by an ASIC vendor. That IC may belong to the application-specific standard product (ASSP) portion of the semiconductor market. ASSPs are sold to many different system vendors, and often may be purchased as standard parts from a catalog, unlike ASICs.

Custom ICs are typically optimized for a specific process technology and take significantly more time to design than ASICs, but can achieve higher performance and low power by higher quality design and use of techniques that are generally available to ASICs. For example, custom designers may design logic gates at the transistor-level to provide implementations that are optimal for specific design; whereas an ASIC designer is limited by what is available in the standard cell library.

2.2.2 FPGA

FPGA is an abbreviation for Field Programmable Gate Arrays. FPGA is the product of further development on the basis of PAL(Programmable Array Logic), GAL(Generic Array Logic), CPLD (Complex Programmable Logic Device) and other programmable devices. As a semi-custom circuit in the field of ASIC, it not only solves the shortcomings of custom circuit, but also overcomes the shortcoming of the limited number of gates of the original programmable devices.

FPGA adopts the concept of Logic Cell Array (LCA), which includes three parts: configurable logic block (CLB), input and output block (IOB), and internal connection. FPGA is a programmable device. Compared with the traditional logic circuits and the gate arrays (such as PAL, GAL and CPLD devices), FPGA has different structures. It uses a small look-up table ($16 \times 1RAM$) to implement combinatorial logic. Each look-up table is connected to the input of a D flip-flop, which then drives other logic circuits or drives I/O ports, thus constituting the basic logic units that can realize both combinatorial logic and sequential logic. The logic of FPGA is to load programming data into internal static storage units to work. The value stored in the memory cell determines the logical function of the logic unit and the connection between each module and I/O. It ultimately determines the function that FPGA can achieve.

FPGA allows unlimited programming. System designers can connect blocks of logic inside FPGA via editable connections as needed, as if a circuit test board had been placed in a chip. The logic blocks and connections of a finished FPGA can be changed according to the designer, so FPGA can complete the required logic functions.

The speed of FPGA is generally slower than that of ASIC, and the area of realizing the same function is larger than that of ASIC. But they also have many advantages, such as quick production, modification to correct errors in the program and cheaper costs. Vendors may also offer cheap but poorly edited FPGA. Because these chips have poor editable capabilities, the development of these designs is done on an ordinary FPGA, and then the design is transferred to a chip similar to ASIC. Another approach is to use CPLD (Complex Programmable Logic Device).

As early as the mid-1980s, FPGA had taken root in PLD devices. CPLD and FPGA included a large number of editable logic units. The density of CPLD logic gates ranged from thousands to tens of thousands of logic units, while FPGA typically ranged from tens of thousands to millions. The circuits which are designed with hardware description language (Verilog or VHDL) can be easily synthesized and distributed, and can be quickly burned to FPGA for testing. It is the mainstream of modern IC design verification technology. These editable elements can be used to implement basic logic gates (such as AND, OR, XOR, NOT) or more complex combinatorial functions such as decoders or mathematical equations. In most FPGA, these editable components also contain memory elements such as triggers (flip-flop) or other more complete memory blocks.

Hardware description languages including VHDL, Verilog, System C and Handle-C are frequently used for FPGA programming. VHDL and Verilog are matured as industry standards. HDLs provided by many vendors can offer simulation and synthesis tools. Behavioral, RTL and structural levels of description can be used inter-changeably in these languages. System C is a C++ based library used for modeling system level behavior. As the base language is C++, software processes can be more easily modeled than in a more traditional HDL. Synthesis tools for System C are emerging, but do not approach the maturity of VHDL or Verilog synthesis products. Handel-C is also a relatively new product in comparison to VHDL or Verilog. Handel-C follows the Communicating Sequential Process (CSP) model. Handel-C requires the designer to explicitly delineate parallel processing blocks within a process. It includes intrinsic for inter-process communication, as does System C 2.0.

2.3 Classification of IC

IC can be classified by two methods. One method is that it can be divided into analog circuit, digital circuit and digital-analog mixed signal circuit. The other is it can be classified by its scale of integration.

2.3.1 Classified by Signal Mode

2.3.1.1 Digital IC

Digital IC is a very small wafer which contains millions of logic gates, flip-flops, memories, multiplexers, adders and other circuits. Here the data registers and counters are discussed as an example to illustrate the matters needing attention in digital IC design.

The simplest type of register is a data register, which is used for the temporary storage of a "word" of data. In its simplest form, it consists of a set of N D flip-flops, all sharing a common clock. All of the digits in the N bit data word are connected to the data register by an N-line "data bus".

Another common form of register used in computers and in many other types of logic circuits is a shift register. It is simply a set of flip-flops (usually D latches or RS flip-flops) connected together so that the output of one becomes the input of the next, and so on in series. It is called a shift register because the data are shifted through the register by one bit position on each clock pulse.

On the leading edge of the first clock pulse, the signal on the data input is latched in the first flip-flop. On the leading edge of the next clock pulse, the contents of the first flip-flop is stored in the second flip-flop, and the signal which is presented at the data input is stored in the first flip-flop, etc. Because the data are entered one bit at a time, this is called a serial-in shift register. Since there is only one output, and data leave the shift register one bit at a time, then it is also a serial out shift register. (Shift registers are named by their methods of input and output, either serial or parallel.) Parallel input can be provided through the use of the preset and clear inputs to the flip-flop. The parallel loading of the flip-flop can be synchronous (i.e., occurs with the clock pulse) or asynchronous (independent of the clock pulse) depending on the design of the shift register. Parallel output can be obtained from the outputs of each flip-flop.

Communication between a computer and a peripheral device is usually done serially, while computation in the computer itself is usually performed with parallel logic circuitry. A shift register can be used to convert information from serial form to parallel form, and vice versa. Many different kinds of shift registers are available, depending upon the degree of sophistication required.

Counters are weighted coding of binary numbers. In a sense, a shift register can be considered a counter based on the unary number system. Unfortunately, a unary counter would require a flip-flop for each number in the counting range. A binary weighted counter, however, requires only flip-flops to count to N. A simple binary weighted counter can be made using T flip-flops. The flip-flops are attached to each other in a way so that the output of one acts as the clock for the next, and so on. In this case, the position of the flip-flop in the chain determines its weight.

A count down counter can be made by connecting the $\overline{Q}$ output to the clock input in the previous counter. By the use of preset and clear inputs, and by gating the output of each T flip-flop with another logic level using AND gates (say logic 0 for counting down, logic 1 for counting up), then a preset-able up-down binary counter can be constructed.

2.3.1.2 Analog IC

Analog ICs are ICs composed of capacitors, resistors and transistors to process analog signals. There are many analog ICs, such as operational amplifiers, analog multipliers, phase-locked loops, power management chips and so on. The main circuits of analog IC are amplifier, filter, feedback circuit, reference circuit, switch capacitor circuit and so on. Here the operational amplifiers are discussed as an example to illustrate the matters need attention in analog IC design.

In order to discuss the ideal parameters of operational amplifiers, we must first define the terms, and then go on to describe what we regard as the ideal values for those terms. At first sight, the specification sheet for an operational amplifier seems to list a large number of values, some in strange units, some interrelated, and often confusing to those unfamiliar with the subject. The approach to such a situation is to be methodical, and take the necessary time to read and understand each definition in the order that it is listed. Without a real appreciation of what each means, the designer is doomed to failure. The objective is to be able to design a circuit from the basis of the published data, and know that it will function as predicted when the prototype is constructed. It is all too easy with linear circuits, which appear relatively simple when compared with today's complex logic arrangements, to ignore detailed performance parameters which can drastically reduce the expected performance.

Let us take a very simple but striking example. Consider a requirement for an amplifier having a voltage gain of 10 at 50kHz driving into a 10kW load. A common low-cost, internally frequency-compensated op amp is chosen; it has the required bandwidth at a closed-loop gain of 10, and it would seem to meet the bill. The device is connected, and it is found to have the correct gain. But it will only produce a few volts output swing when the data clearly shows that the output should be capable of driving to within two or three volts of the supply rails. The designer has forgotten that the maximum output voltage swing is severely limited by frequency, and that the maximum low-frequency output swing becomes limited at about 10kHz. Of course, the information is in fact on the data sheet, but its relevance has not been appreciated. This sort of problem occurs regularly for the inexperienced designer. So the moral is clear: always take the necessary time to write down the full operating requirements before attempting a design. Attention to the detail of the performance specification will always be beneficial. It is suggested the following list of performance details be considered:

1) Closed loop gain accuracy, stability with temperature, time and supply voltage.

2) Power supply requirements, source and load impedances, power dissipation.

3) Input error voltages and bias currents. Input and output resistance, drift with time and temperature.

4) Frequency response, phase shift, output swing, transient response, slew rate, frequency stability, capacitive load driving, overload recovery.

5) Linearity, distortion and noise.

6) Input, output or supply protection required. Input voltage range, common-mode rejection.

7) External offset trimming requirement.

Not all of these terms will be relevant, but it is useful to remember that it is better to consider them initially rather than to be forced into retrospective modifications.

2.3.1.3 Mixed signal IC

Mixed signal ICs are ICs that combine analog and digital circuits. In our daily life, mixed signal circuits are ubiquitous. In general, a chip can perform certain full functions or sub-functions in a large combination, such as, the RF subsystem of a mobile phone, or reading the data path in a DVD player. They usually contain a whole system chip. The A/D converter and the D/A converter are both mixed signal circuits. The analog-to-digital converter converts an analog signal to a digital signal, while an digital-to-analog converter converts a digital signal into an analog signal.

ICs incorporating both digital and analog functions have become increasingly prevalent in the semiconductor industry. Complex digital circuits are now commonly combined with analog circuits as part of the continuing drive toward higher levels of electronic system integration. For example, complex microprocessors are frequently combined with high-performance analog and mixed signal circuits to form "system on chip (SoC)" devices.

System on Chip is an IC that integrates computers or other electronic systems into a single chip. SoC can process digital, analog, mixed and even higher frequency signals. System chips are often used in embedded systems. The integration scale of SoC is very large, generally reaching millions to tens of millions of gates.

Although micro-controllers usually have less than 100kB of random access memory, in fact it is a simple, weakened, single-chip system, and the term "system chip" is often used to refer to more powerful processors. These processors can run some versions of Windows and Linux. More powerful system chips require external memory chips, such as some system chips equipped with flash memory. System chips can often be connected to additional peripherals. The integrated scale of semiconductor devices is required by the system chip. To perform more complex tasks better, some system chips use multiple processor cores.

Mixed signal IC test and measurement have grown into a highly specialized field of electrical engineering. However, test engineering is still a relatively unknown profession compared with IC design engineering. It has become harder to hire and train new engineers to become skilled mixed signal test engineers. It may take one to two years for a mixed signal test engineer to gain enough knowledge and experience to develop adequate test solutions. The slow learning curve for mixed signal test engineers is largely due to the shortage of written materials and university-level courses on the subject of mixed signal testing.

Mixed signal ICs are often designed for specific purposes, but they may also be multipurpose standard components. Mixed signal IC design requires a very high degree of professionalism and careful use of computer-aided design tools. Automated testing of chip completion is also more challenging than general IC. Teradyne and Agilent Technologies are major suppliers of mixed signal chip testing equipment.

2.3.2 Classification based on the Scale of ICs

In this way, ICs can be classified into small scale ICs (SSI), medium scale ICs (MSI), large scale ICs (LSI), very large scale ICs (VLSI) and ultra large scale ICs (ULSI). The differences are shown in Table 2-1.

Table 2-1 The standards for dividing the scale of ICs

classification	digital IC		analog IC
	MOS IC	bipolar IC	
SSI	$<10^2$	<100	<30
MSI	10^2—10^3	100—500	30—100
LSI	10^3—10^5	500—2000	100—300
VLSI	10^5—10^7	>2000	>300
ULSI	10^7—10^9		

2.4 Design of ICs

2.4.1 Design Methods

2.4.1.1 Gate array design

Gate array design technology is a kind of rapid automation design technology for ASIC in the early 1980s. The internal circuit of gate array is an array structure composed of the basic unit layout of a specific structure. The logic unit circuit is realized by the layout of the basic unit. The logic unit library is consist of the basic logic unit circuits. Circuit logic must be implemented by the logic unit library. Gate array design through automatic layout and wiring to complete the layout of the automatic design.

According to the characteristics of the array routing, the gate array can be divided into two types: the wired and unwired channel gate arrays (gate sea). The wired channel gate array is suitable for metal wiring under two layers, and the gate sea is suitable for metal wiring technology with more than three layers, and the cell utilization ratio of the gate sea is high. Unwired arrays are called master chips, which can be serialized according to their size to meet the requirements of circuits of different sizes. The master chip is processed in advance, and the different logic circuits can be realized by changing their wiring.

Gate array design technology has the advantages of short circuit development period, low cost, suitable for many varieties, and small batches of ASIC. With the development of design technology, embedded modular gate array is developed, that is, embedded random access memory module, fixed memory module, and even some IP core modules in the gate array structure.

2.4.1.2 Full-custom design

Full-custom design is to complete the layout design of the whole chip according to the function required by the user's need. Full customization design is a design method from circuit simulation to complete layout design and test design of the whole chip. A lot of work of this kind of design should be done manually. Full Custom design is not convenient to utilize the achievement of the existing circuit directly. The design period is longer and the cost is higher. The artificial layout can make more reasonable use of the chip area, so the integration degree is higher. The artificial layout can make more reasonable use of the chip area, so the integration degree is higher. It is generally smaller than the semi-custom circuit chip area. The performance is relatively ideal. It is very suitable for the development of products with large batches.

2.4.1.3 Structured design

Structured design is a fuzzy expression with different meanings in different expressions. It is a new form of industrial production. Structured ASIC is a kind of customized chip with various characteristics between FPGA and ASIC. It is superior to FPGA, in mass production cost, logic gate utilization ratio, power consumption, and efficiency speed, but not as excellent as pure ASIC. At the same time, it also has the function of programmable logic of FPGA, and accelerates the design speed and modification flexibility of the chip. Thus, the chip can be completed and put into the market more quickly, and the cost of modifying the circuit in the future can be reduced.

At present, the well-known companies of structured ASIC are Altera, AMI Semiconductor, Fujitsu, NEC, ChipX (also known as Chip Express), LSI Logic and so on, among which LSI Logic has announced that it is fading out of structured ASIC business in April 2006.

2.4.2 Design Tools

To assist our customers and reduce design cycle time, we have developed and provided an extensive set of design libraries and tools at zero cost to customers. It includes Cadence, Synopsys, Mentor Graphics and so on.

2.4.2.1 Introduction of cadence

At Cadence, we've pledged ourselves to total customer satisfaction. In fact, commitment to customer success is one of our core company values. In support of this promise, Cadence president and CEO Ray Bingham created the Worldwide Customer Advocacy (WCA) organization. To ensure that every customer is satisfied with the Cadence experience, WCA was recently expanded to include Education Services and Customer Support. These changes allow WCA to focus on the primary mission of establishing a customer-centric corporate culture-one that fosters a universal commitment to customer needs, stronger customer relationships, and superior business value.

With revenues of $1.43 billion in 2001, Cadence Design Systems, Inc. is the world's leading developer of electronic design automation (EDA) products and services. EDA software is a form of computer-aided design and engineering (CAD/CAE) software specifically geared towards automating the design of electronic systems and ICs. The company's products and services are used by companies in the computer, communication, and consumer electronics industries to design and develop

ICs and electronic systems, including semiconductors, computer systems and peripherals, telecommunications and networking equipment, wireless products, automotive electronics, and various other electronic products.

Cadence Design Systems, Inc. was incorporated in June 1988 as the new company resulting from the merger of CAD software companies, ECAD Inc. and SDA Systems, Inc. Computer-aided design of ICs was a rapidly growing field when Cadence started operations. As chip manufacturers tried to fit increasingly more tiny transistors on each chip, the complex layout of the chip's design and its verification came to depend on design automation software.

Cadence continued its strategy of offering software for multiple computer platforms. In addition to versions for Sun Microsystems, Digital Equipment, and Apollo computers, Cadence began to make softwares available for Hewlett-Packard, Sony, and NEC computers in 1989 and 1990. Meanwhile, its competitors, Mentor Graphics Corporation, Daisy Systems Corporation, and Valid Logic Systems, Inc., continued to bundle software with computer hardware for turnkey systems. Software portability was already common for various applications, but it did not become common for CAD software until the early 1990s.

Cadence soon became the world's leading supplier of IC, or chip, design software. In 1989, Cadence took 15.4% market share, ahead of Seiko Instrument and Electronics with 11.5% (mostly in Japan), and Mentor Graphics Corporation with 8.4%, according to the market research firm Dataquest, Inc. One year later, Dataquest put Cadence's market share at 44.2%.

2.4.2.2　Introduction of Synopsys

At Synopsys we talked a lot about commitment, and what it took to be a greatenduring company. A key element of Synopsys' success is our strong set of core values: Integrity, Execution Excellence and Leadership. We adhere to these principles because they are right, and achieve results. Our first commitment is to work for our customers. Synopsys is dedicated to enabling its customers to design amazing chips by delivering the most complete semiconductor design solution through a combination of software, services, and intellectual property. Customer satisfaction is Synopsys' most critical measure. We track success through "one chip at a time."

Synopsys, Inc. is the leading developer of software used in designing semiconductors, a field known as electronic design automation or EDA. The company's products help engineers to develop and test the design of ICs before production, allowing designers to achieve the optimal standards of cost, power consumption, and size in the electronic "chips" used in innumerable products. Synopsys sells its products to semiconductors, computers, communications, consumer electronics and aerospace manufacturers. The company operates more than 60 sales and Research and Development Offices in North America, Europe, Japan, Israel, and the Pacific Rim.

Aart de Geus, who led a work team at General Electric Microelectronics Center, developed a set of ideas for a new software technology called Synthesis. With Synthesis, engineers would be able to "write" the functionality of a circuit in the computer language, rather than describing it in terms of individual gates. The software would automatically create the logic synthesis, saving design time and freeing engineers to focus on creative design solutions rather than manual implementation. Synthesis would create circuit designs from hardware languages (such as VHDL and Verilog), supporting the new generation of EDA technology, hardware language design automation (HLDA).

In 1986, Aart de Geus and several other engineers received support from General Electric and formed Optimal Solutions, Inc., dedicating to the development of Synthesis software. After building the initial prototype, the company was relocated in Mountain View, California, renaming itself Synopsys (SYNthesis OPtimization SYStems). In 1987, EDA entrepreneur Harvey Jones became the president and CEO, leaving Daisy Systems, where he had been a president, CEO, and cofounder.

From 1986 until 1990, the company focused on becoming "The Synthesis Company," as well as changing the methodology of modern electronic design. Synopsys quickly jumped to the forefront of top-down design companies, launching an era that would be defined by top-down design.

Synopsys chased Cadence for years, at last passing the company in late 2003. One of the decisive moments in the final leg of the pursuit was Synopsys' acquisition of Avant! In 2002, which fleshed out Aart de Geus's capabilities in the EDA's most lucrative market segment: developing softwares for creating the smallest and most complex chips.

2.4.2.3 Introduction of mentor graphics

Mentor Graphics has the broadest industry portfolio of best-in-class products and is the only EDA Company with an embedded software solution.

The business plan presented to the financiers revealed that the team had found a market of full potential and all the other necessary ingredients for success. The Computer Aided Engineering (CAE) business is a superset of the CAD/CAM industry. We can define computer aided engineering products as equipment and software which enable an engineer to improve productivity, work quality and consistency. The plan noted rapid growth in CAD/CAM and to an even greater degree in the emerging special purpose workstation environment. Driving forces in the field were identified as a shortage of both engineers and trained technicians and the push for productivity and quality in basic manufacturing industries.

Traditional EDA tools for physical design and verification have reached limits due to greater manufacturing process variability and the growing size and complexity of designs that take advantage of the latest nanometer scaling. With the advent of new process technologies, the handoff between IC layout and manufacturing has changed from a simple check to a multi-step process where the layout design must be enhanced to ensure efficient manufacturing. This presents a host of challenges related to manufacturing process effects, photo-lithography and data volumes. This also achieves a cost-effective yield of finished chips from each wafer.

The Mentor Graphics Tessent product suite is a comprehensive silicon test and yields analysis platform that addresses the challenges of manufacturing test, debug, and yield ramp for today's SoCs. Built on the foundation of the best-in-class solutions for each test discipline, Tessent brings them together in a powerful test flow that ensures total chip coverage.

Tanner EDA by Mentor Graphics is an integrated, affordable and intuitive product suite for the design, layout and verification of analog/mixed signal (AMS) and Micro-Electro-Mechanical Systems (MEMS) ICs. Used by more than 25,000 designers around the world, Tanner tools support an end-to-end flow: top-down; mixed signal design capture and simulation; synthesis with DFT support; physical design; place-and-route; and "sign-off ready" timing analysis to tape-out.

2.5 Low-Power Design of CMOS IC

2.5.1 Introduction

Arguably, invention of the transistor was a giant leap forward for low-power microelectronics that has remained unequal to date, even by the virtual torrent of developments it forbore. Operation of a vacuum tube required several hundred volts of anode voltage and a few watts of power. In comparison, the transistor required only milliwatts of power. Since the invention of the transistor, decades ago, through the years leading to the 1990s, power dissipation, though not entirely ignored, was of little concern. The greater emphasis was on performance and miniaturization. Applications are applied to a battery-pocket calculator, hearing aids, implantable pacemakers, portable military equipment used by individual soldier and most important wrist-watches-drove low-power electronics. In all such applications, it is important to prolong the battery life as much as possible. And now, with the growing trend towards portable computing and wireless communication, power dissipation has become one of the most critical factors in the continued development of the microelectronics technology.

2.5.2 Sources of Power Dissipation

There are three sources of power dissipation in a digital complementary metal-oxide-semiconductor (CMOS) circuit. The first source is the logic transitions. As the "nodes" in a digital CMOS circuit transition back and forth between the two logic levels, the parasitic capacitances are charged and discharged. Current flows through the channel resistance of the transistors, and electrical energy is converted into heat and dissipated away. As suggested by this informal description, this component of power dissipation is proportional to the supply voltage, node voltage swing, and the average switched capacitance per cycle. The second source of power dissipation is the short-circuit currents that flow directly from supply to ground when the n-subnetwork and the p-subnetwork of a CMOS gate both conduct simultaneously. The third and the last source of dissipation is the leakage current that flows when the input(s) to, and therefore the outputs of, a gate is not changing. This is called static dissipation. In current day technology the magnitude of leakage current is low and usually neglected. As the supply voltage is being scaled down to reduce dynamic power, however, MOS field-effect transistors (MOSFETs) with low threshold voltages have to be used. The lower the threshold voltage, the lower the degree to which MOSFETs in the logic gates are turned off and the higher is the standby leakage current.

The power dissipation attributable to the three sources described above can be influenced at different levels of the overall design process.

Since the dominant component of power dissipation in CMOS circuits varies as the square of the supply voltage, significant savings in power dissipation can be obtained from operation at a reduced supply voltage. If the supply voltage is reduced while the threshold voltages stay the same, that will reduce the noise margins are reduced. To improve noise margins, the threshold voltages need to be made smaller as well. However, the subthreshold leakage current increases exponentially when the threshold voltage is reduced. The higher static dissipation may offset the reduction in transitions component of the dissipation. Hence the devices need to be designed to have threshold voltages that maximize the net reduction in the dissipation.

The transitions component of the dissipation also depends on the frequency or the probability of occurrence of the transitions. If a high probability of transitions is assumed and correspondingly low supply and threshold voltages chosen, to reduce the transitions component of the power dissipation and provide acceptable noise margins, respectively, the increase in the static dissipation may be large. As the supply voltage is reduced, the power-delay product of CMOS circuits decreases and the delays increase monotonically. Hence, while it is desirable to use the lowest possible supply voltage, in practice, only as low a supply voltage can be used as corresponds to a delay that can be compensated by other means, and steps can be taken to retain the system level throughput at the desired level.

One way of influencing the delay of a CMOS circuit is to change the channel-width to channel-length ratio of the devices in the circuit. The power-delay product for an inverter driving another inverter through an interconnect of certain length varies with the width to length ratio of the devices. If the interconnect capacitance is insignificant, the power-delay product initially decreases and then increases when the width-to-length ratio is increased and the supply voltage is reduced to keep the delay constant. Hence, there exists a combination of the supply voltage and the width-to-length ratio that is optimal from the power-delay product consideration.

The way to assure that the system level throughput does not degrade as supply voltage is reduced by exploiting parallelism and pipelining. Hence as the supply voltage is reduced, the degree of parallelism or the number of stages of pipelining is increased to compensate for the increased delay. But the latency increases. Overhead control circuitry also has to be added. As such circuitry itself consumes power, there exists a point beyond which power increases, instead of decreasing, Even so, great reductions in power dissipation by factors as large as 10, have been shown to be obtainable theoretically as well as in practice.

At the logic level, automatic tools can be used to locally transform the circuit and select realizations for its pieces from a precharacterized library so as to reduce transitions and parasitic capacitance at circuit nodes and therefor circuit power dissipation. At a higher level, various structural choices exist for realizing any given logic functions; for example, for an adder, one can select one of ripply-carry, carry-look-ahead or carry-select realizations.

In synchronous circuits, even when the output computed by a block of combinatorial logic is not required, the block keeps computing its output from observed input every clock cycle. In order to save power, entire execution units comprising of combinatorial logic and their state registers can be put in a stand-by mode by disabling the clock and/or powering down the unit. Special circuitry is required to detect and power-down unused units and power them up again when they need to be used.

The rate of increase in the total amount of memory per chip as well as rate of increase in the memory requirement of new applications has more than kept pace with the rate of reduction in power dissipation per bit of memory. As a result, in spite of the tremendous reductions in power dissipation obtained from each generation of memory to the next, in many applications, the major portion of the instantaneous peak-power dissipation occurs in the memory.

In case of dynamic RAM memory, the most effective way to reduce power of any memory size is to lower the voltage and increase the effective capacitance to maintain sufficient charge in the cell. The new array organizations introduced recently present many possibilities of lowering the power. A far greater challenge at this point is to get to the next generation of memory chips with a capacity of 4Gb. At the 1Gb generation, there is no design margin remaining after the required bit area is subtracted from the available bit area. The implications are that it may not be possible to implement the capacitor-and-transistor cell for the 4Gb memory in the conventional folded bit line architecture.

2.5.3 Synthesis for Low Power

In order to meet functionality, performance, cost-efficiency and other requirements, automatic synthesis tools for IC design have become indispensable. The designs are described at the Register Transfer or higher level. A technology-independent logic/architectural level realization is generated next. The logic level realization is then mapped to a specific technology library. Finally the technology-mapped circuits are optimized to ensure that all requirements have been met. The process includes multiple steps because the problem of realizing the specified functionality with gates in a given library to meet all the requirements is too complex to be solved in one step.

Tools have been developed to carry out each of those steps automatically. In the beginning, the synthesis tools attempted to reduce area alone and did not consider delay. Next came improved algorithms that, in addition to reducing the area, ensured that the maximum delay through the circuit did not increase or at least remained within the specified maximum bound. With the exponential growth in the number of gates that can be accommodated on a chip continuing unabated, the area has become less of a concern and is increasingly being substituted for performance and power dissipation.

(1) Behavioral level transforms

Because of many degrees of available freedom, the behavioral level has the potential of producing a big improvement in power dissipation. At this level, since operations haven't been assigned, the execution time and hardware allocation have not been performed yet. A design point with given power dissipation, delay and area can be found if one exists in the design space. Hence, there is a great need to systematically explore the design space. Traditionally, behavioral synthesis has targeted optimization of the amount of hardware resources required and optimization of the average number of clock cycles per task required to perform a given set of tasks. In the past few years, a handful of studies have been carried out to examine the efficacy of behavioral level techniques to reduce power dissipation.

(2) Logic level optimizations for low power

When, to begin with, a design is described at the behavior level, first behavioral synthesis and then logic synthesis are performed. Unfortunately for the proponents of behavioral synthesis, it has not been widely accepted. Most designs are initially described at the Register Transfer or logic level. Hence logic synthesis tools are often the first tools in the synthesis flow. The data path sections of the designs usually have regular and pre-optimized structures. As a result, logic optimization of data path sections is usually limited to binding the logic gates to the cells or gates in the technology library. The control sections, on the other hand, require technology-independent optimizations to achieve small area and delay.

For finite-state machines (FSMs) found in the control logic, state assignment has to be carried out. State assignment strongly influences the random logic associated with the FSMs. Hence, proper state assignment for low-power dissipation is required. For the random logic, which can be looked upon as a set of Boolean functions to map the inputs into outputs, logic synthesis involves determining the common set of sub-functions between the given Boolean equations in order to optimize that area. However, it has been observed that minimum area is not always associated with minimum power dissipation for CMOS circuit.

2.6 Microelectromechanical Systems

Microelectromechanical systems (MEMS) integrate mechanical and electrical components and have feature sizes ranging from micrometers to millimeters. They may be fabricated using methods similar to those used to construct ICs and they have the potential of providing significant cost advantages when MEMS were fabricated. Their size also makes it possible to integrate them into a wide range of systems. Feature sizes may be made with size on the order of the wavelength of light, making them attractive for many optical applications. Microsensors (e.g., accelerometers for automobile crash detection and pressure sensors for biomedical applications) and microactuators (e.g., for moving arrays of micromirrors in projection systems) are examples of commercial applications of MEMS.

2.6.1 Introduction

MEMS are micrometer-scale devices that integrate electrical and mechanical elements. They have been used in diverse applications, from display technologies to sensor systems to optical networks. MEMS are attractive for many applications because of their small size and weight, which allow systems to be miniaturized.

To further explore the challenges and opportunities of MEMS, this article describes some fundamental technologies associated with MEMS design and operation. First, the advantages and challenges of MEMS are discussed. Then manufacturing processes are presented, followed by the methods used to generate micro-scale forces and motions. Finally, several example MEMS applications are highlighted.

2.6.2 Advantages and Challenges of MEMS

The small size of MEMS is attractive for many applications because feature sizes are typically as small as 1 micrometer or less. Hence, for optical applications, features may be made with size in the order of the wavelength of light. Their small size also allows applications to be developed which would otherwise be impossible. For example, micromechanical switches fabricated as part of a communications circuit allow phase shifting and signal switching at speeds that would be impossible to achieve using macro-scale switches. To illustrate the scale of a typical microsystem, Figure 2-5 shows a micromachined mirror assembly next to a spider mite. The mirror assembly is about 100 micrometers wide, and it is dwarfed by the spider mite. The mechanical devices surrounding the mirror allow it to be positioned accurately as part of an optical network.

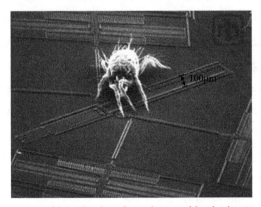

Figure 2-5　A spider mite dwarfs a micromachined mirror assembly

Other advantages include the on-chip integration of electromechanical systems and the circuitry used to control them, allowing further miniaturization. Furthermore, many MEMS fabrication technologies allow parallel fabrication of thousands of systems by leveraging the parallel fabrication techniques of the IC industry. This may lead to a reduction in the manufacturing cost and improvement in reliability.

Like any technology, microsystems present some challenges. Because micro mechanisms operate at a size scale far below that of typical mechanical devices, surface forces such as adhesion and friction may dominate over other forces in the system, leading to failure of the device. Careful device design, fabrication, and testing are required to reduce these effects. Their small size also makes it difficult to interact mechanically with MEMS components. In many MEMS, electrical or optical signals are used to interface with, provide power to, and control the device. In many MEMS, electrical or optical signals are used to interface to provide power and control the device, rather than the manual, hydraulic, or pneumatic control typically seen in macro-scale mechanical devices. In addition, packaging of MEMS components has often presented a challenge because each device must be packaged in a way that keeps the components clean and free from contamination, while also allowing mechanical motion and, in many cases, interaction with the environment. For example, a MEMS pressure sensor requires a package that exposes the sensor to the ambient pressure while protecting the electronic circuitry from dust or other particles. Finally, while parallel fabrication techniques can reduce the manufacturing cost of many units, MEMS development may be more costly because few units are produced at a time using complex and expensive fabrication equipment.

2.6.3 Fabrication Technologies

A large variety of fabrication methods have been employed for MEMS. However, many of these methods may be broadly described under three headings: surface micromachining, bulk micromachining and molding.

2.6.3.1 Bulk micromachining

Bulk micromachining, the oldest of the micromachining technologies, is accomplished by removing material from a substrate to create holes, cavities, channels or other desired shapes. Early bulk micromachining was accomplished using isotropic or anisotropic wet etching of silicon or glass substrates. In particular, several chemicals, such as KOH (potassium hydroxide) or TMAH (tetramethylammonium hydroxide) etch a silicon substrate preferentially depending on the crystalline planes in the direction of etching. The etch rate for these chemicals is tens to hundreds of time faster in the [100] crystalline plane compared to the [111] plane. This effect has been used to create a wide variety of features using simple wet etching.

Another common technique of bulk micromachining uses a deep reactive ion etch (DRIE) plasma etcher. Using this technique, silicon, as well as some other materials, can be etched very quickly and very anisotropically, making possible very thick structures with small widths. For example, the mirrors in Figure 2-6 were deep etched from a titanium substrate. These mirrors are flatter and more rigid than mirrors made using surface micromachining.

Figure 2-6　Titanium mirrors bulk micromachined using deep etching of a titanium substrate.

Another important technology frequently used with bulk micromachining is wafer bonding. After etching the desired parts in several different substrates, the substrates may be bonded together to create systems incorporating several parts. In some cases, the wafers are bonded using an adhesive, but this method often has poor accuracy in spacing and alignment between bonded parts. Silicon and glass wafers can be bonded more accurately by applying high temperatures (about 1,000℃) to a clean interface between the surfaces. In many cases, the bond generated in this way is as strong as the intrinsic strength of glass or silicon. If high temperatures are not desired, anodic bonding can be used. In this case, the wafers are pushed together at a lower temperature (about 400 ℃), and then a large voltage (about 1,000V) is placed between the wafers. This generates a high electric field at the interface, accelerating bonding even at the lower temperature. Wafer bonding has also been explored as a way to create a vacuum package enclosing MEMS parts.

2.6.3.2　Surface micromachining

Surface micromachining is one of the most common technologies used to manufacture MEMS devices. In surface micromachining, films are deposited on a substrate and patterned. Micromechanical devices are created with photolithography. The films normally alternate between structural and sacrificial layers, with the MEMS parts being made from the structural layers. The sacrificial layers serve to support the structural components during fabrication. After the structural layers are patterned, the sacrificial material is removed. Wet chemical etching is often used. The result is freestanding MEMS parts that can move relative to the fixed substrate.

Most early surface micromachining used polycrystalline silicon (polysilicon) as the structural layers and an oxide of silicon as the sacrificial material. However, as surface micromachining has further developed, numerous other materials have been used. Depending on the desired application, MEMS developers have used metals, oxides and nitrides of silicon, and even polymers for both structural and sacrificial films. Several foundry processes have also been developed to allow users to design their own MEMS and have them fabricated with the foundry's fabrication facilities. For example, Figure 2-7 shows a surface micromachined electrostatic drive manufactured with the SUMMiT foundry process developed at Sandia National Laboratories. The device is made from two layers of released polysilicon with a layer of polysilicon. The fingers of the electrostatic drive are about 35 micrometer long, and each layer is about 2 micrometer thick. The gaps (between the substrate and the first layer, and between the two layers) are determined by the thickness of the sacrificial oxide film—2 micrometers.

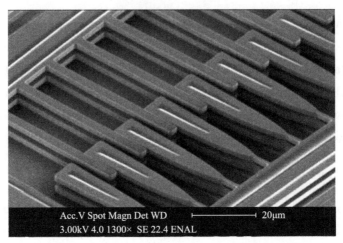

Figure 2-7　An electrostatic comb drive fabricated with SUMMiT, a surface-micromachining foundry process

2.6.3.3　Molding

Finally, MEMS parts are often made by creating a mold, which may then be filled to create the desired part. Molds have been made from a variety of polymers, including some types of photoresist, as well as metal and deep-etched silicon wafers. Photolithography is normally used to define the mold pattern. If metal parts are desired, the mold may be filled by electroplating. Polymer parts may be created by pouring or pressing the precursor into the mold. After the part has been molded, it may be removed from the mold by either etching the mold away, or, if the mold is to be used again, by peeling away the mold.

Micro-molding was first performed in Germany, where it was called LIGA, an acronym for the German words lithography, electroplating and molding. The original LIGA process required an X-ray source to fully expose thick layers of photosensitive material, but many molding techniques have since been developed that use visible or ultraviolet light sources. However, because of this history, many molding processes are still referred to as LIGA or LIGA-like processes. Figure 2-8 shows a copper waveguide fabricated with true LIGA technology. Because of its high thickness and exact specifications, the waveguide has very low losses, especially compared to other micromachined waveguides.

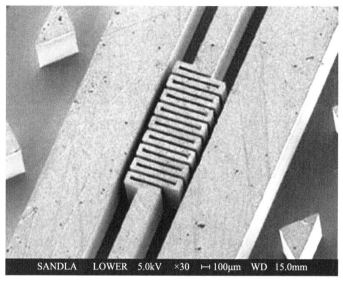

Figure 2-8　An extremely low-loss coplanar waveguide fabricated using LIGA

2.7　Summary

This chapter mainly introduces the related knowledge of IC. The main contents in this chapter include the history of the development of ICs, the definition of IC and the difference between IC and separated circuit, the classification of IC, the design methods and design tools of IC, and the low power design of CMOS IC MEMS are described. All of the above contents can let the readers have a comprehensive understanding of the IC.

New Words and Expressions

integrated circuit	集成电路
microprocessor	微处理器
semiconductor	半导体
rectification	整流
carrier	载流子
vacuum tube	真空管

digital	数字的
electrode	电极
track	轨迹，路线
keep track of	记录，与……保持联系
cellular	蜂窝状，多孔的
amplify	放大，扩大
transistor	晶体管
resistor	电阻
capacitor	电容
diode	二极管
bipolar junction transistor	双极型晶体管
Moore's law	摩尔定律
n-channel	n 沟道
p-channel	p 沟道
complementary-MOS	CMOS，互补 MOS
application-specific IC	(ASIC)专用集成电路
memory	存储器
system on chip	片上系统
hardware description language (HDL)	硬件描述语言
application-specific standard product (ASSP)	专用标准电路
low power	低功耗
field-programmable gate array (FPGA)	现场可编程门阵列
flip-flop	触发器
analog	模拟的
digital	数字的
quartz-crystal	石英晶体
oscillator	振荡器
serializer-deserializer	串行器-解调器
simulation	模拟
synthesis	综合
multiplexer	多路复用器
synchronous	同步的，同时发生的

asynchronous	异步的
counter	计数器
sensor	传感器
power management circuit	电源管理电路
operational amplifier	运算放大器
bandwidth	带宽
transient response	建立时间
slew rate	转换速率
computer aided design (CAD)	计算机辅助设计
wafer	晶圆，芯片
gate array design	门阵列设计方法
full-custom design	全定制设计方法
standard-cell library	标准单元库
leakage current	漏电流
static dissipation	静态功耗
dynamic power	动态功耗
threshold voltage	阈值电压
noise margin	噪声容限
behavioral level	行为级
logiclevel	逻辑级
finite-state machine (FSM)	有限状态机
Boolean function	布尔函数
Microelectromechanical system	微机械系统

Exercises

1. 请将下述词语翻译成英文

集成电路	无线传输	双极型晶体管	带宽
摩尔定律	金属连线	逻辑门	n沟道场效应管
设计工具	硬件描述语言	模数转换器	闭环增益
短路电流	漏电流	门阵列设计方法	阈值电压

2. 请将下述词语翻译成中文

ASIC	FPGA	SOC	RTL
EDA tool	ASSP	FPGA	shift register
digital-analog converter	filp-flop	data-bus	phase shift
metal-oxide semiconductor	power management circuit		output swing
mixed signal IC	VLSI	EDA	noise margin

3. 请将下述短文翻译成中文

1) In 1947, Bardeen, Brattain, and Shockley discovered a Bipolar Junction transistor and the Modern Age began. It was considered a revolution. Being small, fast, reliable and effective, it quickly replaced the vacuum tube.

2) In 1958, at Texas Instruments, Jack Kilby was probably most famous for his invention of the IC (as shown in Figure 2-2), for which he received the Nobel Prize in Physics in the year 2000.

3) An electric circuit is made from different electrical components such as transistors, resistors, capacitors and diodes that are connected to each other in different ways. These components have different behaviors.

4) IC uses the semiconductor production process, with many transistors and resistors, capacitors and other components in a small silicon wafer. The components are combined into a complete electronic circuit with the wiring method of multilayer.

5) FPGA is an abbreviation for Field Programmable Gate Arrays. FPGA is the product of further development on the basis of PAL (Programmable Array Logic), GAL (Generic Array Logic), CPLD (Complex Programmable Logic Device) and other programmable devices. As a semi-custom circuit in the field of ASIC, it not only solves the shortcomings of custom circuit, but also overcomes the shortcoming of the limited number of gates of the original programmable devices.

6) Digital IC is a very small wafer which contains millions of logic gates, flip-flops, memories, multiplexers, adders and other circuits. Here the data registers and counters are discussed as an example to illustrate the matters need attention in digital IC design.

7) Mixed signal ICs are often designed for specific purposes, but may also be multipurpose standard components. Mixed signal IC design requires a very high degree

of professionalism and careful use of computer-aided design tools. Automated testing of chip completion is also more challenging than general IC. Teradyne and Agilent Technologies are major suppliers of mixed signal chip testing equipment.

8) Cadence soon became the world's leading supplier of IC, or chip, design software. In 1989, Cadence had a 15.4% market share, ahead of Seiko Instrument and Electronics with 11.5% (mostly in Japan), and Mentor Graphics Corporation with 8.4%, according to the market research firm Dataquest, Inc. One year later, Dataquest put Cadence's market share at 44.2%.

9) Synopsys sells its products to semiconductor, computer, communications, consumer electronics and aerospace manufacturers. The company operates more than 60 sales and Research and Development Offices in North America, Europe, Japan, Israel and the Pacific Rim.

10) Tanner EDA by Mentor Graphics is an integrated, affordable and intuitive product suite for the design, layout and verification of analog/mixed signal (AMS) and Micro-Electro-Mechanical Systems (MEMS) ICs. Used by more than 25,000 designers around the world, Tanner tools support an end-to-end flow: top-down, mixed signal design capture and simulation; synthesis with DFT support; physical design; place-and-route; and "sign-off ready" timing analysis to tape-out.

11) And now, with the growing trend towards portable computing and wireless communication, power dissipation has become one of the most critical factors in the continued development of the microelectronics technology.

12) Since the dominant component of power dissipation in CMOS circuits varies as the square of the supply voltage, significant savings in power dissipation can be obtained from operation at a reduced supply voltage.

References

An Introduction to Cadences. https://community.cadence.com/.
Field-programmable gate array. https://en.wikipedia.org/wiki/Field programmable_gate_array.
Introduction of Mentor Graphics. https://www.mentor.com/products/mechanical/history.
Introduction of Synopsys. https://news.synopsys.com/index.php?s=20295&item=122657.
Introduction to Microelectromechanical Systems (MEMS). https://en.wikipedia.org/wiki/ Microelectromechanical_systems.

Rabaey Jan M. 1996. Digital Integrated Circuits. New Jersey: Prentic-Hall, Inc.

Roy K, Prasad S. 2000. Low-Power CMOS VLSI Circuit Design. New York: A Wiley Interscience Publication.

Zhang A H, Chen W P, Li T. 2003. English in Electronic Science and Technology. Harbin: Harbin Institute of Technology Press.

Chapter 3 Electric Circuit

3.1 Diode and Rectifier Circuit

3.1.1 Introduction

If you want to get known of the electronic technology, there are several electronic components that you must know about, such as diodes, transistors, resistors, capacitors and inductors. In this chapter, I will give you a brief introduction to the diode, and bring you into the wonderful world of Electronics. Here, you will find that the original complex electronic technology is also very interesting. Gradually you will find the joy of it.

In the electronic component family, there is one kind of element that only allows current to flow in one direction and has two electrodes. It is the cornerstone of the modern electronic industry—the diode.

3.1.2 The History of Diode

(1) Vacuum electron diode

Early diodes were mainly divided into "Cat's Whisker Crystals" and "Thermionic Valves". In 1904, British physicist Fleming invented the world's first electronic diode in accordance with the Edison effect. It relies on the cathode to emit hot electrons to the anode to achieve conduction. If the positive and negative poles of the power supply and diode are connected in reverse, the diode will not conduct electricity. Therefore, it is a kind of electronic devices that can only conduct current in one direction.

In the early stage, the electronic diode (Figure 3-1) had the problems of large volume, high power consumption and easy-breaking, which led to the invention of crystal diode.

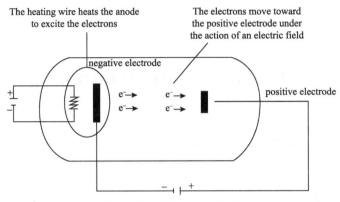

Figure 3-1 Electronic diode

(2) Crystal diode

Crystal diode is the most common diode invented by Americans in 1947. Since it is mostly constructed with semiconductor materials such as silicon or germanium structures, it is also called the semiconductor diode. The electronic device has a pn junction and two terminals formed inside. According to the direction of the applied voltage, it has the conductivity of unidirectional current. Semiconductor diodes are one of the earliest semiconductor devices, which are widely used in almost all of the electronic circuits, and play an important role in the circuit.

3.1.3 The Structure of Diode

In order to understand the structure and working principle of the crystal diode, we need to make clear the concept of semiconductor, p-type semiconductor and n-type semiconductor.

In nature we can divide the material into conductors, insulators and semiconductors, based on the electrical conductivity of them. Copper and aluminum are common conductors. Rubber and plastic are common insulators. What is a semiconductor? The conductivity of semiconductors is between a conductor and an insulator, and the common semiconductor material is silicon (Si) and germanium (Ge). To this point, these two materials are made into commonly used semiconductors: silicon tubes and germanium tubes. It is clear that the diode or the transistor is made of silicon or germanium as a substrate.

Semiconductors are heat-sensitive, photosensitive and doable. According to the heat sensitivity and photosensitivity, we can make the temperature sensor and light sensor with semiconductor; in addition, we can manufacture p-type semiconductors and n-type semiconductors by utilizing the semiconductor doping characteristics (the conductivity of semiconductor can be improved when semiconductors are mixed with a small amount of impurity elements). Thus, we can make the p-type semiconductor and n-type semiconductor.

We call an entirely impurity-free semiconductor as an intrinsic semiconductor. p-type semiconductors are formed by doping with trace elements of boron (B) or gallium (Ga) in intrinsic semiconductors such as silicon or germanium. If it is doped with pentavalent element phosphorus (P) in intrinsic semiconductors, it is called n-type semiconductor. If a trace amount of trivalent element boron (B) is added to a part of a silicon or germanium semiconductor, a p-type semiconductor is made. If a trace amount of the pentavalent element phosphorus (P) is added to another part of the semiconductor, making this part an n-type semiconductor is made. A pn junction is formed at the junction of the p-type and n-type semiconductors. A pn junction is called a diode. The lead in the p-type area is called the anode (also called the positive electrode), and the lead wire in the n-type area is called the cathode (also called the negative electrode) so that a semiconductor diode is constructed (Figure 3-2).

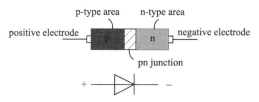

Figure 3-2   Structure of pn junction

3.1.4 Diode Conduction Characteristics

As one of the most commonly used electronic components, the most obvious characteristic of the diode is the unidirectional conductivity. It means that the diode is a device that controls an electric current so that it can only flow in one direction. The volt-ampere characteristic shows in Figure 3-3.

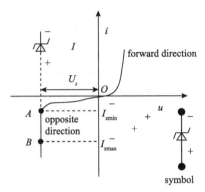

Figure 3-3   Volt-ampere characteristic of diode

(1) Forward characteristic

In the electronic circuit, the positive electrode of the diode is connected at the high potential end and the negative electrode is connected with the diode at the low potential end. This connection is called forward bias. It means that when the forward voltage applied across the diode is low, the diode barely conducts electricity, and the forward current flowing through the diode is very weak. Only when the forward voltage reaches a certain value (this value is called the "threshold voltage", also known as the "dead voltage"; the germanium tube is about 0.1V and the silicon tube is about 0.5V) can make diode conduct. After the diode at both ends of the voltage is essentially unchanged (approximately 0.3V for a germanium tube and approximately 0.7V for a silicon tube), it is referred to as the diode's "forward voltage drop".

(2) Reverse characteristic

In the electronic circuit, the positive electrode of the diode is connected at the low potential end and the negative electrode is connected with the diode at the high potential end, there is almost no current flowing through the diode. At this time, the diode is in the cut-off state. This connection is called reverse bias. Only weak reverse current flows through the diode at this time, which is called leakage current. When the reverse voltage increases to a certain value, the reverse current will increase dramatically. The diode will lose the unidirectional conductivity and this state is called the breakdown of the diode.

3.1.5   Typical Applications of Diode

We can use the one-way conductivity of the diode to build the rectifier circuit, detection circuit, voltage regulator circuit and modulation circuit. The invention of the diode and other components led to the birth of our rich and colorful electronic information world.

(1) Rectifier diode

Most diodes have the current direction of what we usually call "Rectifying". As shown in Figure 3-4, the bridge rectifier circuit can realize full-wave signal rectification.

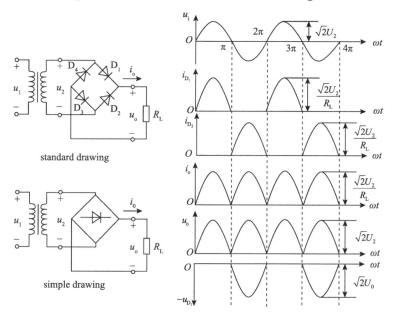

Figure 3-4 Bridge rectifier circuit

(2) Detector diode

Detector diodes are used to detect the low frequency signal which is superimposed on the high frequency carrier wave.

(3) Switching diode

Switching diode is a kind of semiconductor diode. It is a kind of diode for "on" and "off" in circuit. Switching diode has the characteristics of high speed, small size, long service life and high reliability. It is widely used in electronic equipment switch circuit, detection circuit, high frequency and pulse rectifier circuit and automatic control circuit.

(4) Voltage stabilizing diode

Voltage stabilizing diode is a diode which uses the pn junction to reverse breakdown state. (It is a phenomenon that the current can be changed over a wide range but the voltage is almost constant.)

(5) Light emitting diode

Light emitting diode is referred to as LED. It is composed of gallium (Ga), arsenic (As), phosphorus (P), nitrogen (N) and other compounds. A lot of characteristics of light emitting diode are not comparable to ordinary light emitting devices. The main features are safety, high efficiency, environmental protection, long life, fast response, small size and strong structure. Therefore, the light emitting diode is a light source which meets the requirements of the green lighting. At present, it is widely used in various electronic products, such as light source, optical fiber communication light source, instruments, as well as lighting. For example, it is used in LCD TV, computer screen, media player MP3, MP4, as well as mobile phones and other display screens as a light emitting diode screen back light source.

3.2 History and Typical Application Circuit of the BJT

The invention of the BJT provided opportunities for the third industrial revolution. In just half a century, it was widely applied to the invention of the computer. The human civilization was rapidly moving from the electric era to the era of automation. The wide use of transistors in computer technology has made an indelible contribution to the development of people's life, and the characteristics of the transistor brought itself broad development prospects. So the research on development and application of the transistor not only has very important academic significance but also has a great social significance.

3.2.1 The History of BJT

On December 23, 1947, Murray Hill of New Jersey Baer Lab, Dr. Bardeen, Dr. Brighton and Dr. Shockley worked on their experiment intensely and systematically. They were working on the experiment that used the semiconductor crystal to amplify the sound signal in a conductor circuit. They were amazed to find that a small amount of current passed through the devices they invented could actually control the large currents that flow through the other part. This is amplification effect. That device is transistor which is the epoch-making achievement in technology. It was invented on the eve of Christmas and had such a great impact on people's lives in the future, so it was called "Christmas gifts to the world". In addition, the three scientists won the Nobel Prize in physics in 1956.

Transistors promote and bring a "Solid revolution", and then promote the semiconductor electronics industry on a global scale. As the main component, first and foremost, it was used in the communication tools and made great economic benefits. With the transistor completely changed the structure of the electronic circuit. The IC and the large scale IC emerged, which made the manufacture of high precision devices such as high-speed electronic computers become a reality.

3.2.2 The Structure and Operational Principle

BJT, just as its name implies, has three electrodes. If you make two closely spaced pn junctions in one semiconductor chips, the common electrode is called base (represented by b) of the BJT transistor, the other two electrodes are respectively called collector (represented by c) and emitter (represented by e). On account of different compound modes, there are two different kinds of BJT—one type is npn and the other type is pnp. The two kinds are shown in Figure 3-5(a) and (b).

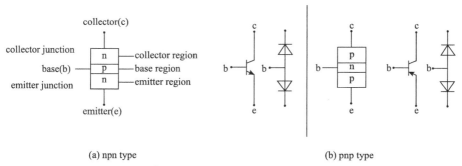

Figure 3-5　BJT structure diagram

The basic function of BJT is amplification, which can amplify weak electric signals (shortened as signal, such as voltage, current, power) to desired value without distortion, and partly transform the DC power supply into output signals which change regularly according to the input signals and contain vast amount of energy. Of course, this transformation still keeps to the principle of conservation of energy, and it just transforms the energy of power source into signal energy. The essence of amplifying circuit is a kind of energy conversion device to control the conversion of larger energy with smaller energy.

The critical parameter of BJT is current amplification factor β. When injecting weak electrical currents into BJT transistor's base, current which is five times the strength of the original one will be produced on the collector, called collector current. Collector current varies with the changing in the base current and a very small change in the base current can cause a large change in the collector current, which is named the amplification of the BJT.

Let's take an example of basic common-emitter amplifier circuit shown in Figure 3-6. It shows the amplification of the BJT (the BJT was symbolized with VT). First let's give an introduction of the concept of "quiescent operating point": what is called "quiescent". It means the relationship of base current (I_{bQ}), collector current (I_{cQ}), B_{ceQ}, and U_{ceQ} is only under the action of the DC signal U_{cc} when the AC signal u_i is not applied. We can see a static circuit as Figure 3-7 shows through analyzing the electro circuit in Figure 3-6.

$$I_b = \frac{U_{cc} - U_{be}}{R_b}$$

$$I_b \approx \frac{U_{cc}}{R_b}$$

$$I_c \approx \beta I_b$$

$$U_{ce} = U_{cc} - I_c R_c$$

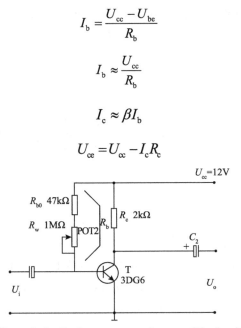

Figure 3-6　Basic common-emitter amplify circuit

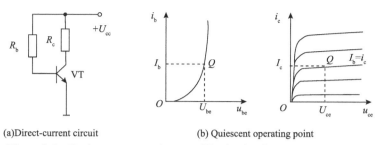

(a)Direct-current circuit　　　　　(b) Quiescent operating point

Figure 3-7　Basic common-emitter amplify circuits direct-current circuit and quiescent operating point

Appropriate setting and proper adjustments within a limited numerical range can situate the "quiescent operating point" around the position shown in Figure 3-7(b) simultaneously, the BJT is in the amplifier region, so we can say that the BJT is operating under saturated amplification; but if it is over-amplifiable, the BJT may be in cut-off region; on the contrary, the BJT will be operating under saturation region and come out of amplification function. Therefore, we need to adjust to keep the BJT amplifiable in order to ensure the amplification function of the electrocircuit.

3.2.3　The Typical Application of the BJT

(1) The most simple audio amplifier circuit

Figure 3-8 shows us a schematic diagram of the most concise audio amplifying circuit composed of the basic common-emitter amplifier. By analyzing the schematic, we can know that to keep the signal without distortion when the circuit reaches the maximum output, we can adjust the size of the resistor values. Of course, greater output power can be available if we can reduce the resistor values. The structure of this circuit is extremely simple, but the effects of power supply voltage to the bias current can be relatively large caused by the fixed bias.

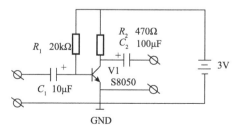

Figure 3-8　Schematic of audio amplifier circuit

(2) BJT for electronic switches

Analyzing the schematic of electronic switching circuit in Figure 3-9, and combining with the introduction of BJT's operation principle we have done before, we can see: when the input signal $u_i=0$, the BJT work in the cut-off area, the output U_o is high-level U_{cc}; and if the input U_i is high, making the BJT work in a saturated and conductive situation, then the output U_o is no longer U_{cc}, but a low level after the voltage drop across R_c subtracted. Thus, this feature of BJT is like the characteristics of a switch. The output will be characterized by the high level when the input signal is extremely small or zero; and when the input signal is relatively large, the output will be characterized by the low level, as if the switch is closed. That's why we call this circuit for the electronic switching circuit.

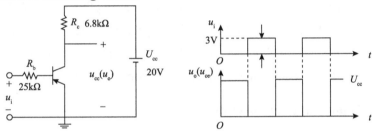

(a) Schematic of electronic switching circuit (b) Input, output signal waveform

Figure 3-9 Electronic switching circuit of BJT

In addition to common functions such as amplifiers and electronic switches, the BJT can also be used with other components to constitute the oscillator and other typical practicable circuits.

3.2.4 The Development Trend of Transistor

The BJT has unparalleled development advantages due to it's simple structure and wide application. Personally, microelectronics technology will be widely used in the near future and the BJT will have a bright prospects. Nowadays, we can see them from the automotive electronic products in the automotive electronics industry. With the development of integrated electronic technology, especially the coming era of high-tech civilian, the BJT will be more and more integrated into our lives in medical and public construction to show its development potential.

3.3 Signal Operation and Processing

In the electronic circuit, how to achieve the mathematical operation of analog signals? Operational amplifiers were originally designed to perform mathematical operations such as addition, subtraction, differentiation and integration. The first amplifier designed with a vacuum tube was completed around 1930 and can perform addition and subtraction operations. Because of its ability to achieve all kinds of mathematical operations, operational amplifier comes to be the basic blocks of analog computers.

With the development of technology, the performance of operational amplifiers, whether they are made with transistors, vacuum tubes, discrete components, or IC components, is very close to ideal operational amplifiers. It's commonly used in precision instruments and automatic test equipment and other instruments. This characteristic of operational amplifier is more used than the arithmetic calculation in circuit design.

3.3.1 History and Current Situation of Operational Amplifiers

In the late 1960s, Fairchild Semiconductor produced the first widely used IC operational amplifier, model μA709, designed by Bob Widlar. But the μA709 was quickly replaced by a new product, the μA741, which was better, stabler, and easier to use. The μA741 operational amplifier has not been replaced for decades and became a unique symbol in the history of the microelectronics' industry. Many manufacturers of ICs are still producing μA741 operational amplifiers, and the model of component will be added "μA741" to distinguish. But in fact, there are still many better operational amplifiers than the μA741 appeared, such as field-effect transistors (FETs) in the 1970s or the metal oxide semiconductor field effect-transistor (MOSFET) in the early 1980s. These components can be used directly in the μA741's circuit architecture to get better performance.

Operational amplifier specifications are usually strictly limited. Packaging and power supply requirements have also been standardized. The operational amplifier can perform a variety of analog signal processing tasks, with only a small number of external components. Nowadays, while the price of standard or general-purpose operational amplifiers fall below \$1 due to large demand and production, special-purpose operational amplifiers can still cost more than a hundred times the price of a universal version.

3.3.2 Working Principle of Operational Amplifiers

Operational amplifier, shortened as "op-amp", is an important multi-terminal circuit component. The symbol of operational amplifier is shown in Figure 3-10.

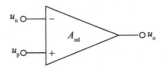

Figure 3-10 Symbol of the operational amplifier circuit

Figure 3-10 shows the symbol of operational amplifier. It has two inputs (labeled "+" and "−") and an output. The input labeled "+" is called the "non-inverting input" and the voltage/current is marked as u_p/i_p. The input labeled "−" is called "inverting input" and its voltage/current is marked as u_n/i_n. The output is labeled u_o. Its amplification is denoted as "A_{od}". The operational amplifier can be connected to a single power supply or a dual power supply, as shown in Figure 3-11.

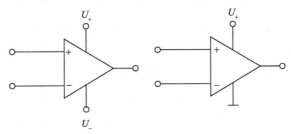

Figure 3-11 The two power supply that operational amplifier can be accessed

The relationship between the output voltage u_o and the input voltage of the operational amplifier $u_p - u_n$ can be approximately described in Figure 3-12.

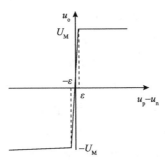

Figure 3-12 Curve of voltage transmission characteristics of operational amplifier

When $|u_p - u_n| = \varepsilon \approx 0$, the output u_o varies linearly with the value of the input $u_p - u_n$, and its slope is the amplification A_{od}. Since the A_{od} value is large, this curve is very steep. This operational state of the operational amplifier is usually called linear region. In this region, when requirements of the power supply is met, the relationship between the output and the input of the operational amplifier can be expressed as

$$u_o = A_{od} u_i = A_{od}(u_p - u_n) \tag{3-1}$$

When $|u_p - u_n| > \varepsilon$, The output voltage is nearly saturated, $u_o = \pm U_M$, U_M is a finite voltage value, its value is slightly lower than the value of the power supply voltage and we usually refer to this state as saturation region.

3.3.3 Operational Characteristics and Typical Application Circuit of the Operational Amplifier

(1) Working in linear region

If operational amplifier works in the linear region, it will show some interesting features. We can get a lot of unique functions by using these features flexibly. To sum up, there are two characteristics:

1) The amplification of an operational amplifier is infinite.

2) The input resistance of the operational amplifier is infinite and the output resistance is zero.

Now let's take a brief look at what conclusions can be drawn from these two features.

First of all, the amplification A_{od} is very large when the operational amplifier works in the linear region and we can ideally think that the A_{od} is approximately infinite. Due to the power supply voltage limit, the output voltage u_o must be a finite value, and its absolute value does not exceed the absolute value of the power supply voltage. Therefore, $(u_p - u_n)$ in the equation must be approximately the reciprocal of the infinity, which is about zero, so u_p is approximately equal to u_n, just like u_p and u_n are shortened. We call this state as "virtual short circuit". That is, $u_p \approx u_n$.

Secondly, the input resistance R_i of operational amplifier is infinite, and the equivalent input current $i_p = i_n = (u_p - u_n)/R_i$. Because R_i is infinite and $u_p \approx u_n$. Thus, the input current $i_p = i_n \approx 0$, just like the wire of the non-inverting input and the inverting input are broken. However, they are not really broken, so that the phenomenon is "virtual circuit", that is, $i_p = i_n \approx 0$.

Example of typical Circuit 1: Adder circuit

The adder is the most basic logical operation unit inside the computer. So we use the operational amplifier to achieve an addition operation circuit. The circuit is shown in Figure 3-13.

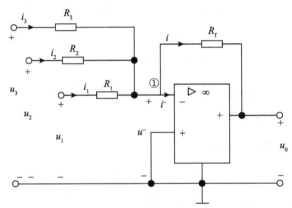

Figure 3-13 The circuit of adder with operational amplifier

The operational amplifier has negative feedback from the output to the inverting input. Therefore it works in the linear region, so $u_p = u_n = 0 \text{ V}$, and

$$i_1 = \frac{u_1}{R_1} \tag{3-2}$$

$$i_2 = \frac{u_2}{R_2} \tag{3-3}$$

$$i_3 = \frac{u_3}{R_3} \tag{3-4}$$

Due to virtual circuit, $i_p = i_n \approx 0$, so

$$i = i_1 + i_2 + i_3 \tag{3-5}$$

The output

$$u_o = -i \times R_f \tag{3-6}$$

We can get the output u_o from Eq. (3-2)—Eq. (3-6),

$$u_o = -\left(\frac{u_1}{R_1} + \frac{u_2}{R_2} + \frac{u_3}{R_3}\right)R_f$$

If $R_f = R_1 = R_2 = R_3$, then

$$u_o = -(u_1 + u_2 + u_3) \tag{3-7}$$

From the above equations, we can know that the output of this circuit is the inverse of the sum of the three signals from the input. So this circuit can achieve the addition operation.

Example of typical Circuit 2: Integrating circuit

We can know the operational amplifier works in the linear region from Figure 3-14.

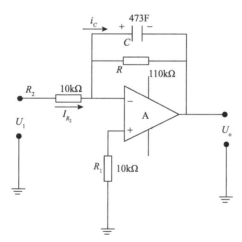

Figure 3-14 Integrating circuit

So $u_p = u_n$ and $i_p = i_n \approx 0$. Then

$$i_p = 0 \Rightarrow u_p = 0, \quad u_n = 0$$

We can get the output

$$u_o = -\frac{1}{C}\int_{t_1}^{t_2} i_C dt + U_C\big|_{t_1} \tag{3-8}$$

$$i_C = i_{R_2} = u_i / R_2 \tag{3-9}$$

We can get the following equation with (3-9) and (3-8):

$$u_o = -\frac{1}{R_1C}\int_{t_1}^{t_2} U_i dt + U_C\big|_{t_1} \tag{3-10}$$

$U_C\big|_{t_1}$ is the voltage across C when time is t_1. If the voltage across C_1 at time t_1 is zero, the above formula turns to be Eq. (3-11)

$$u_o = -\frac{1}{R_1C}\int_{t_1}^{t_2} U_i dt \tag{3-11}$$

If we input the square wave input to the integral circuit with the amplitude of U_{in} shown in Figure 3-15(a), when $U_1 = -1V$, the output voltage U_o increases linearly over time, when $U_1 = 1V$ and the output voltage U_o decreases linearly over time. The square wave transforms into the amplitude of triangular wave through the integral circuit as shown in Figure 3-15 (b).

$$U_o = \frac{1}{R_1C}\int_0^{\frac{T}{4}} U_{in} dt = \frac{U_{in}T}{4R_1C} \tag{3-12}$$

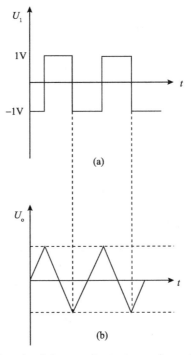

Figure 3-15　Integrating circuit input and output waveform when input square wave

From the above equation, we can analyze that the amplitude of the output triangular wave is proportional to the amplitude of the input signal and the period T, inversely proportional to the time. The error of the integrating circuit is mainly caused by the capacitor of operational amplifier which is not ideal.

(2) Working in saturation region

When the operational amplifier works in the saturation region, the output will no longer change linearly with the input. Instead, the output is high level U_m when $u_p - u_n > 0$. On the contrary, when $u_p - u_n < 0$, the output is low level $-U_m$. Therefore, the "virtual short circuit" conclusion no longer holds. Because the input resistance tends to infinity, the principle "virtual circuit" is still true and $i_p = i_n \approx 0$.

If the hysteresis comparator and integrator are connected end to end, they can form a positive-feedback and closed-loop system as shown in Figure 3-16. The square wave outputted from the comparator "A_1" is integrated by the integrator "A_2" to obtain a triangular wave. Then triangular wave triggers comparator to automatically flip to form a square wave, so you can constitute a square and triangle wave generator. The square and triangular wave generator's output waveform is shown in Figure 3-17. Using the integral circuit composed of the operational amplifier can achieve constant current charging. Thus, the triangle wave can be greatly improved.

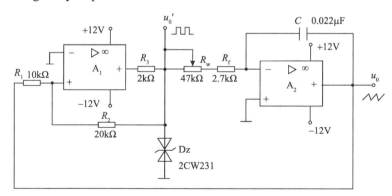

Figure 3-16　Square and triangle wave generator with operational amplifier

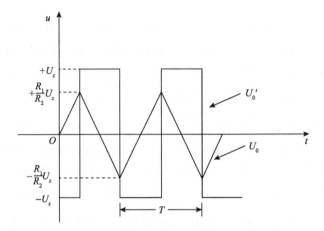

Figure 3-17   Output waveform of square-triangular wave generator

Circuit oscillation frequency

$$f_o = \frac{R_2}{4R_1(R_f + R_w)C_f}$$

Square wave amplitude

$$U'_{om} = \pm U_z$$

Triangular wave amplitude

$$U_{om} = \frac{R_1}{R_2}U_z$$

Adjustment R_w can change the oscillation frequency, and the change of the ratio $\frac{R_1}{R_2}$ can adjust the amplitude of the triangular wave.

$$f_o = \frac{1}{4R_1(R_f + R_w)R_1}\int_0^{\frac{T}{4}} U_{im}dt = \frac{U_{im}T}{4R_1C_f} \tag{3-13}$$

$$U'_{om} = \pm U_z \tag{3-14}$$

$$U_{om} = \pm \frac{R_1}{R_2}U_z \tag{3-15}$$

3.4 Design of DC Regulated Power Supply

3.4.1 Introduction

People nowadays are greatly affected by the convenience of electronic equipment. Each electronic device has a common circuit—power circuit. From a device as big as a supercomputer to a gadget as small as a pocket calculator, all the electronic equipment must be working properly with the support of power circuit. It's obvious that the power supply circuit is the basis of all electronic equipment, and there will not have such a wide variety of electronic equipment without power circuit. For the electronic equipment, the requirement of the power circuit is to provide continuous stable electrical energy that can also meet the load requirements. And usually a stable supply of DC power is required. That is the DC regulated power supply which provides stable DC power.

3.4.2 Design Objective

Firstly, we use 220V AC power supply transformers through the dual 15V transformator to output dual 15V AC. Secondly rectified the AC voltage to DC voltage through the diode rectifier bridge, then filtered by the filtering capacitor in order to output smoother voltage. Lastly, with three-terminal fixed integrated regulator LM7805, LM7905, LM7812, LM7912 as the core composition, we finally get the DC regulator circuit. Because the integrated voltage regulator has high reliability and it can also help to improve the accuracy of the regulator, reduce volume and reduce weight, the output DC is basically free from the influence of grid voltage fluctuation and load resistance. It can achieve high enough stability.

Specific parameters:

1) The input voltage is AC 220V, 50Hz. The power output voltage is ±12V. The required voltage error is less than 0.2V.

2) Ripple voltage $\Delta V_{op\text{-}p} \leqslant 5mV$.

3) The regulator coefficient $S_r \leqslant 5\%$.

4) Maximum output current $I = 1A$.

5) Output resistance R_o : 50 to 100Ω.

3.4.3 Circuit Composition and Implementation

Low-power regulated power supply is usually composed of four parts of the circuit: transformer, rectifier, filter and regulator. Transformer only need to use the transformator to decrease the amplitude of input signal moderately. The rest of the functional circuit is shown as follows.

3.4.3.1 Rectifier circuit

Rectifier circuit uses unidirectional conductivity of the diode converting the AC voltage into one-way pulsating DC voltage. Commonly used rectifier circuit are half-wave rectifier, full-wave rectifier and bridge rectifier. In this section we utilize bridge rectifier circuit, the circuit structure, input and output waveform shown in Figure 3-18.

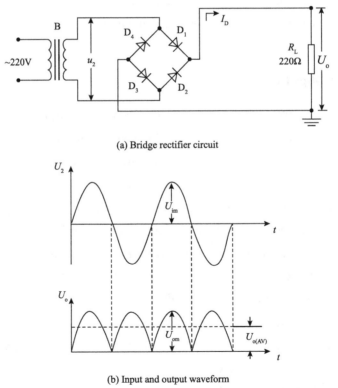

(a) Bridge rectifier circuit

(b) Input and output waveform

Figure 3-18　Bridge rectifier circuit, input and output waveform

One of the indicators that measure the performance of the rectifier circuit is the average of the output voltage $U_{o(AV)}$, as shown in Eq. (3-16). (Where: U_2 is the valid value when sinusoidal AC voltage is input to the rectifier circuit.)

$$U_{o(AV)} = \frac{1}{2\pi} \int_0^{2\pi} U_2 \mathrm{d}\omega t \qquad (3\text{-}16)$$

For bridge rectifier circuits,

$$U_{o(AV)} = \frac{2\sqrt{2}}{\pi} U_2 \approx 0.9 U_2 \qquad (3\text{-}17)$$

In order to facilitate the measurement practically, we often use the ripple factor γ to describe the size of the rectified output pulse, which is defined as the ratio of the total valid values to the mean in the output U_o AC component, that is

$$\gamma = \frac{\tilde{U}_o}{U_{o(AV)}} \qquad (3\text{-}18)$$

3.4.3.2 Filter circuit

Rectification just turns AC into a one-way pulsating DC in order to get a smooth DC needs filter circuit. Capacitance filtering is commonly used in low-power DC power supply. The waveform of the output through capacitor filter after bridge rectifier is shown in Figure 3-19 (a).

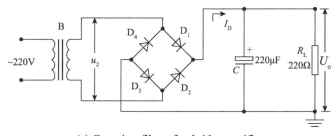

(a) Capacitor filter after bridge rectifier

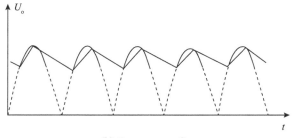

(b) Output waveform

Figure 3-19　Capacitor filter after bridge rectifier and output waveform

The ripple coefficient S is related to the $R_L C$ time constant. For bridge rectifier and capacitor filter, it can generally be approximated as

$$S = \frac{1}{\frac{4R_L C}{T} - 1}$$

where T is the period of AC voltage of the grid. Obviously, when R_L and T are constant, the larger C is, the smaller the ripple coefficient is.

3.4.3.3　Voltage regulator circuit

Although the ripple factor is greatly reduced after filtering, the output voltage is not stable enough. Mainly, when the load current or grid fluctuates, the output voltage will change. So we add voltage regulator measures. At present, integrated voltage regulator circuit is convenient to use and has stable performance and low price, which has basically replaced the discrete components of the regulator circuit. The design uses the most extensive LW7800 series integrated regulator for regulator circuit.

LM7800 series is three-terminal integrated voltage regulator with fixed positive voltage output and the last two digits in the model imply the voltage value of output. Such as LM7805, the output is +5V, W7815, the output is +15V and the rest may be deduced by analogy. The maximum output current of this series is 1.5A. The minimum voltage difference between the input and output is required to be 2V. Figure 3-20 shows the conformation and pin array of a plastic LM7800 three-terminal regulator. "1" feet is for the input, "2" feet is connected to the common side (ground), and "3" feet is for the output.

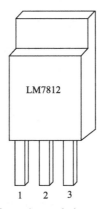

Figure 3-20　Conformation and pin array of LM7800 series

Finally, the DC regulated power supply we designed is shown in Figure 3-21.

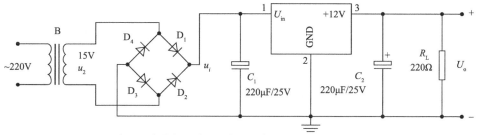

Figure 3-21　The DC regulated power supply

3.4.3.4　The main technical indicators of the regulator circuit

(1) Regulator coefficient S_r

The regulator coefficient reflects the effect of grid voltage fluctuation on output voltage. It is defined as the ratio of the relative change in the output voltage $\dfrac{\Delta U_o}{U_o}$ to the relative change in the input voltage $\dfrac{\Delta U_i}{U_i}$ when the load is fixed, that is

$$S_r = \left.\frac{\Delta U_o / U_o}{\Delta U_i / U_i}\right|_{R_L=\text{constant}} \tag{3-19}$$

where U_i is DC voltage after rectifier filtering.

As we often take the power grid voltage fluctuation ±10% as a limit condition in engineering, so as long as we measure out the DC voltage U_i after rectifier filter and DC power supply output voltage U_o corresponding to power grid voltage of 198V, 220V and 242V, can we calculate the regulator coefficient S_r.

(2) Voltage regulator circuit internal resistance R_o

Voltage regulator circuit internal resistance is defined as the effect of load current's change ΔI_o on the output voltage when voltage regulator circuit input voltage U_i is constant, that is,

$$R_o = \left.\frac{\Delta U_o}{\Delta I_o}\right|_{U_i=\text{constant}} \tag{3-20}$$

The internal resistance of regulator circuit reflects the load characteristics of regulator circuit. As long as U_i is constant, measure out U_o when there is no load ($R_L = \infty$) and output voltage U_{oL} when load current is specified, then use Eq. (3-21) to get R_o.

$$R_o = \frac{\Delta U_o}{\Delta I_o} = \frac{U_o - U_{oL}}{U_{oL}} R_L \tag{3-21}$$

(3) Ripple factor γ

The ripple factor of regulator circuit is the ratio of the ripple voltage $\tilde{U}_o$ output by the regulator circuit to the output DC voltage U_o, that is

$$\gamma = \frac{\tilde{U}_o}{U_o} \tag{3-22}$$

3.4.4 Equipment for the Experiment

1) Oscilloscope, one;
2) AC millivoltmeter, one;
3) Voltage regulator, one;
4) Analog circuit experiment box, one;
5) Digital multimeter, one.

New Words and Expressions

vacuum electron diode	真空电子二极管
crystal diode	晶体二极管
semiconductor diode	半导体二极管
pn junction	pn 结
terminal	引出端
intrinsic semiconductor	本征半导体
p-type semiconductor	p 型半导体
n-type semiconductor	n 型半导体
unidirectional conductivity	单向导电性

forward bias	正向偏置
reverse bias	反向偏置
rectifier diode	整流二极管
detector diode	检波二极管
switching diode	开关二极管
voltage stabilizing diode	稳压二极管
light emitting diode	发光二极管
bipolar junction transistor	双极型晶体管
triode transistor	晶体三极管
large scale IC	大规模集成电路
triode	三极管
collector	集电极
base	基极
emitter	发射极
current amplification factor	电流放大系数
quiescent operating point	静态工作点
amplifier region	放大区
cut-off region	截止区
saturation region	饱和区
the basic common-emitter amplifier	基本共射极放大器
electronic switching circuit	电子开关电路
integrated electronic technology	集成电子技术
operational amplifier	运算放大器
AC and DC amplifier	交直流放大器
active filter	有源滤波器
oscillator	振荡器
voltage comparator	电压比较器
transistor	晶体管
vacuum tube	真空管
discrete	分立式元件
IC	集成电路
field-effect transistor	场效应晶体管

metal oxide semiconductor field effect transistor	金属氧化物半导体场效应晶体管
digital oscilloscope	数字示波器
voltage transmission characteristic	电压传输特性
non-inverting input	同相输入端
inverting input	反相输入端
linear region	线性区
saturation region	饱和区
amplification	放大倍数
virtual short circuit	虚短路
virtual circuit	虚断路
integrating circuit	积分电路
triangle wave	三角波
square and triangle wave	方波
hysteresis comparator	迟滞比较器
positive-feedback	正反馈
closed-loop	闭环
amplitude	幅值
DC regulated power supply	直流稳压电源
integrated regulator	集成稳压器
ripple voltage	纹波电压
regulator coefficient	稳压系数
half-wave rectifier	半波整流
full-wave rectifier	全波整流
one-way pulsating DC	单向脉动直流
sinusoidal AC voltage	正弦交流电压
smooth DC	平滑直流
capacitor filter	电容滤波
ripple coefficient	脉动系数
regulator circuit	稳压电路
pin array	管脚排列图
voltage regulator circuit internal resistance	稳压电路内阻

Exercises

1. 请将下述词语翻译成英文

真空电子二极管　　半导体二极管　　p 型半导体　　单向导电性
正向偏置　　　　　检波二极管　　　稳压二极管　　发光二极管
双极型晶体管　　　三极管　　　　　集电极　　　　基极
发射极　　　　　　静态工作点　　　截止区　　　　振荡器
基本共射极放大器　交直流放大器　　真空管　　　　分立式元件
同相输入端　　　　线性区　　　　　放大倍数　　　虚短路
积分电路　　　　　迟滞比较器　　　闭环　　　　　幅值
直流稳压电源　　　纹波电压　　　　稳压系数　　　全波整流
电容滤波　　　　　正弦交流电压　　脉动系数　　　管脚排列图

2. 请将下述词语翻译成中文

crystal diode　　　　　　　　pn junction　　　　　　　　　terminal
intrinsic semiconductor　　　n-type semiconductor　　　　 reverse bias
rectifier diode　　　　　　　switching diode　　　　　　　triode transistor
amplifier region　　　　　　 saturation region　　　　　　active filter
operational amplifier　　　　voltage comparator　　　　　 transistor
virtual circuit　　　　　　　triangle wave　　　　　　　　positive-feedback
integrated regulator　　　　 half-wave rectifier　　　　　saturation region
smooth DC　　　　　　　　　　one-way pulsating DC　　　　 regulator circuit
IC　　　　　　　　　　　　　 field-effect transistor　　　digital oscilloscope
inverting input　　　　　　　current amplification factor
electronic switching circuit　integrated electronic technology
large scale IC　　　　　　　 metal oxide semiconductor field effect transistor
voltage regulator circuit internal resistance
characteristic voltage transmission

3. 请将下述短文翻译成中文

1) In the electronic component family, there is one kind of element that only allows current to flow in one direction and has two electrodes. It is the cornerstone of

the modern electronic industry—the diode.

2) In 1904, British physicist Fleming invented the world's first electronic diode in accordance with the Edison effect. It relies on the cathode to emit hot electrons to the anode to achieve conduction.

3) Crystal diode is the most common diode invented by Americans in 1947. Since it is mostly constructed with semiconductor materials such as silicon or germanium structures, it is also called the semiconductor diode.

4) As one of the most commonly used electronic components, the most obvious characteristic of the diode is the unidirectional conductivity. It means that the diode is a device that controls an electric current so that it can only flow in one direction.

5) In the electronic circuit, the positive electrode of the diode is connected at the high potential end and the negative electrode is connected with the diode at the low potential end. This connection is called forward bias.

6) We can use the one-way conductivity of the diode to build the rectifier circuit, detection circuit, voltage regulator circuit and modulation circuit. The invention of the diode and other components led to the birth of our rich and colorful electronic information world.

7) Voltage stabilizing diode is a diode which uses pn junction to reverse breakdown state. (It is a phenomenon that the current can be changed over a wide range but the voltage is almost constant.)

8) The invention of the BJT provided opportunities for the third industrial revolution. In just half a century, it was widely applied to the invention of the computer.

9) Transistors promote and bring a "Solid revolution", and then promote the semiconductor electronics industry on a global scale. As the main component, first and foremost, it was used in the communication tools and made great economic benefits.

10) Of course, this transformation still keeps to the principle of conservation of energy, and it just transforms the energy of power source into signal energy. The essence of amplifying circuit is a kind of energy conversion device to control the conversion of larger energy with smaller energy.

11) Thus, this feature of BJT is like the characteristics of a switch. The output will be characterized by the high level when the input signal is extremely small or zero; and

when the input signal is relatively large, the output will be characterized by the low level, as if the switch is closed.

12) With the development of integrated electronic technology, especially the coming era of high-tech civilian, the BJT will be more and more integrated into our lives in medical and public construction to show it's development potential.

13) Operational amplifiers were originally designed to perform mathematical operations, such as addition, subtraction, differentiation and integration.

14) It's commonly used in precision instruments and automatic test equipment and other instruments. This characteristic of operational amplifier is more used than the arithmetic calculation in circuit design.

15) Operational amplifier specifications are usually strictly limited. Packaging and power supply requirements have also been standardized. The operational amplifier can perform a variety of analog signal processing tasks, with only a small number of external components.

16) If operational amplifier works in the linear region, it will show some interesting features. We can get a lot of unique functions by using these features flexibly.

17) From the above equation, we can analyze that the amplitude of the output triangular wave is proportional to the amplitude of the input signal and the period T, inversely proportional to the time.

18) From a device as big as a supercomputer to a gadget as small as a pocket calculator, all the electronic equipment must be working properly with the support of power circuit.

19) For the electronic equipment, the requirement of the power circuit is to provide continuous stable electrical energy that can also meet the load requirements. And usually a stable supply of DC power is required.

20) Because the integrated voltage regulator has high reliability and it can also help to improve the accuracy of the regulator, reduce volume and reduce weight, the output DC is basically free from the influence of grid voltage fluctuation and load resistance. It can achieve high enough stability.

21) Rectifier circuit uses unidirectional conductivity of the diode converting the AC voltage into one-way pulsating DC voltage. Commonly used rectifier circuit are

half-wave rectifier, full-wave rectifier and bridge rectifier.

22) Although the ripple factor is greatly reduced after filtering, the output voltage is not stable enough. Mainly, when the load current or grid fluctuates, the output voltage will change. So we add voltage regulator measures.

References

方志烈. 2007. 发光二极管材料与器件的历史、现状和展望. 湖南理工学院学报, (6): 42-46.
傅吉康. 1982. 怎么选用无线电元件. 北京:人民邮电出版社.
黄继昌, 张海贵, 郭继忠, 等. 2000. 实用单元电路及其应用. 北京: 人民邮电出版社.
康华光. 2008. 电子技术基础. 北京: 高等教育出版社.
康占成. 2007. 模拟电路故障诊断方法浅析. 山西大同大学学报: 自然科学版, 23(1): 82-84.
龙晓庆. 2009. 单电源集成运算放大器的应用探讨. 大众科技, (8): 14-15.
莫正康. 1993. 半导体变流技术. 北京: 机械工业出版社.
童诗白, 华成英. 2001. 模拟电子技术基础. 3 版. 北京: 高等教育出版社.
席时达. 2005. 电工技术. 北京: 高等教育出版社.
殷登祥. 2007. 科学技术与社会概论. 广州: 广东教育出版社.
张昊. 2006. 基准电压源和比较器的实现. 哈尔滨: 哈尔滨工业大学.
支传德. 2006. 射频功率放大器线性化和稳定性的分析与设计. 北京: 清华大学.
Thomas L. 2005. Floyd, Electronic Devices . New Jersey: Person Prentice Hall.

Chapter 4　Writing an Academic Paper

4.1　Various Sections of the Academic Paper

An academic paper is for researchers to report their scientific studies and communicate with other researchers. To write up an academic paper, you are usually suggested to start with an outline, in which you make several crucial things clear, such as the topic, purpose, importance, hypotheses and major findings of your research. Following the advice of George M. Whitesides (2004), a good outline should be "a carefully organized and presented set of data, with attendant objectives, hypotheses, and conclusions, rather than an outline of text". In this section, we suppose that you have already prepared your outline, and further give suggestions on how to construct a paper based on it.

　　An academic paper usually begins with ① a title, ② information of the authors, ③ an abstract, and ④ keywords. Its main body contains ⑤ an introduction, ⑥ methods, ⑦ results, ⑧ a discussion and/or conclusion. Last, it also has ⑨ references and sometimes ⑩ acknowledgements. In the following, we will talk about how to write up and edit an academic paper, with emphasis on the writing of the main body. In some cases, all the contexts cannot be put into a limited number of pages. It is not recommended to change font size or line space to squeeze all words into those pages. The passive voice may be written in the first person singular/plural or directly use the active voice. Don't forget to check spelling of words and sentences. A native English speaker will be very helpful to proofread and enhance the quality of writing. It is easy to download an electronic template from journal websites and use it to prepare your manuscript.

4.1.1　Title, Abstract and Keywords

The title, abstract and keywords are helpful to facilitate readers to find the papers online through search engines such as CNKI, Baidu, IEEE explore or Google Scholar.

A good title should be a clear, accurate and concise description of the paper contents, containing essential words or key phrases for the research topic. Journals usually have specific requirements or suggestions for the title. For example, in *Nature*'s guideline for authors, it says that titles do not exceed two lines and do not normally include numbers, acronyms, abbreviations or punctuation. Paper titles should be written in uppercase and lowercase letters, not all uppercase. Avoid writing long formulas with subscripts in the title. Short formulas that identify the elements are fine. Do not use abbreviations in the title unless they are unavoidable.

The abstract provides an overview of the whole paper through briefly summarising the purpose, methods, results and conclusions. It usually has a specific limitation of word numbers. To write an abstract, you can start with a summary containing what you think is important, and then gradually reduce the words by removing unnecessary information. You can repeat the essential words or key phrases in the abstract to attract potential readers. Be sure to define all symbols used in the abstract. Do not cite references in the abstract. Use italics for emphasis. Do not use underline.

Keywords are supplement to the information given in the title, so do not simply use words or phrases from the title as keywords. Good keywords should well represent your research area, experimental material, techniques, etc. You can use search engines to find out what terms are the most common ones in your research field. Enter key words or phrases in an alphabetical order and separated them with commas.

4.1.2 Introduction

"An introduction should announce your topic, provide context and a rationale for your work before stating your research questions and hypothesis. Well-written introductions set the tone for the paper, catch the reader's interest, and communicate the hypothesis or thesis statement."(http://www.wikihow.com/Write-a-Research-Introduction).

Define abbreviations and acronyms the first time they are used in the text, even after they have already been defined in the abstract. You can write a first draft of your introduction by answering the following questions:

1) What is your research topic?

2) What is a brief background introduction of this research topic?

3) Why is this topic important to research society to investigate?

4) What have been known about this topic before you do this study?

5) What are your hypotheses, objectives and research questions?

6) How will this study advance new knowledge or new ways of understanding?

7) What is the contribution of this work?

4.1.3 Methods

In the methods section, you should present enough description of the experimental materials and procedures necessary to allow evaluation and replication of the research and interpretation of the results. If you find it difficult to start your paper, you may put down the methods section first, because you have all the details for conducting the experiment. Number equations consecutively with equation numbers in parentheses flush with the right margin, as in (1). Be sure that the symbols in your equation have been defined before the equation appears or immediately follows. To begin with, you may consider:

1) What experimental materials do you use?

2) Who/what are the subject(s) of your study?

3) What is the design approach of your research?

4) What does procedure you follow?

5) What are special design tricks to report?

It is a good idea to read some published papers in your field to get some idea of what is included in the method section. The language in this section should be particularly accurate, informative and concise.

(1) Accuracy

Instead of: "Wet paper has limitations in impact resistance, toughness and strength."

Write: "The mechanical properties of wet paper might demonstrate limitations in impact resistance, toughness and strength. " (Camciunal et al., 2016)

(2) Informativeness

Instead of: "We seeded the paper and assessed the deposition of the bone mineral."

Write: "We seeded the paper at densities of 0.1×10^6, 0.4×10^6, and 1.6×10^6 cells per sample and assessed the deposition of the bone mineral (calcium phosphate) microscopically by staining the samples with alizarin red. " (Camciunal et al., 2016)

(3) Concision

Instead of: "The paper we used was Whatman filter paper grade 114. The average size of this paper is 25μm."

Write: "We used Whatman filter paper grade 114, which has an average pore size of 25μm." (Camciunal et al., 2016)

Last, it is worth noting that if your detailed methods have been published previously, you should avoid repeating and provide only the reference.

4.1.4 Results

The results should be presented in subsections in texts, tables, figures, formulas, algorithms and so on. You may obtain a lot of data in the experiment, but you can be selective when presenting them in the paper so that your readers can better understand your essential findings. To make a selection, you may ask yourself what are related to the main research questions of your study, and what are the corresponding results. To insert images in a Word file, place the cursor at the insertion point and use Insert | Picture | From File. The proper resolution of your figures depends on the type of figure. Author photographs, color, and grayscale figures should be at least 300dpi. High-quality pictures are very essential to reveal the results of this research. Therefore, the authors should try their best to make these figures easy to understand by others.

Succinctness and objectivity

Instead of: "Figure 1 clearly describes the culture of cells on scaffolds fabricated from paper."

Write: "Figure 1 describes the culture of cells on scaffolds fabricated from paper." (Camciunal et al., 2016)

Do not interpret or discuss the results in this section, unless you are submitting to journals which use joint results/discussion sections, where results are immediately followed by interpretations.

4.1.5 Discussion and Conclusion

The discussion section is to "explain the meaning of the findings and why they are important …" (Swales and Feak, 2004). To begin with, you can briefly summarise the significant results, but don't repeat what you have said in the results section. Then you should interpret and discuss the results. To develop and deepen your discussion, you may ask yourself the following questions:

1) How do the results answer the research questions?
2) Do the data support the hypotheses?
3) Are the results consistent with previous findings?
4) If the results were unexpected, what can be the potential reasons?
5) Is there another way to interpret the results?
6) What further research is needed to solve the remaining questions (if there is any)?
7) How do your results fit into the big picture?
8) What are the advantages and disadvantages of your work in contrast to the state-of-the-art works in the literature?

In the conclusion, you can make a general statement reiterating the purpose of the study, the answers to the research questions and adding its scientific implications, practical application or advice. Although a conclusion may review the main points of the paper, do not replicate the abstract as the conclusion. A conclusion might elaborate on the importance of the work or suggest applications and extensions.

The language used in this section should convey confidence and authority. To achieve that, you are suggested "to use the active voice and the first person pronouns. Accompanied by clarity and succinctness, these tools are the best to convince your readers of your point and your ideas" (Kallestinova, 2011).

Confidence and authority

Instead of: "It is suggested that the paper-based cell culture platform is a valuable system …"

Write: "Our data suggest that the paper-based cell culture platform is a valuable system …"(Camciunal et al., 2016)

4.1.6　References

The references in the running text and in the reference list should conform to the required reference style (e.g., MLA, APA, Chicago and Harvard). Technical papers submitted for publication must advance the state of knowledge and must cite relevant prior work. In general, you should make sure:

All required periodical data (e.g., volume and page numbers, publisher, place of publication, etc.) are included in the reference list.

All references appeared in the text can be found in the reference list.

References deleted in the text during revision are deleted in the reference list.

4.1.7 Acknowledgements

In this section you usually thank the contributors who helped with the experiments, gave suggestions about the protocol or commented on the manuscript. This section can be optional, though in principle you should list all the important contributors who do not meet the criteria for authorship.

4.1.8 Other tips

1) Every journal conforms to a particular set of standards for writing. You should follow the manuscript formatting guidelines of the journal you are submitting to closely when editing your paper.

2) Be aware of the differences between UK spellings and US spellings (e.g., analyse vs. analyse, programme vs. program, colour vs. color, etc.). Follow the required spelling rules if there is any, and keep consistent throughout the whole paper.

3) Authors must convince both peer reviewers and the editors of the scientific and technical merit of the paper; the standards of proof are higher when extraordinary or unexpected results are reported.

4) Because replication is required for scientific progress, papers submitted for publication must provide sufficient information to allow readers to perform similar experiments or calculations and use the reported results. Although not everything need be disclosed, a paper must contain new, useable and fully described information. For example, a specimen's chemical composition need not be reported if the main purpose of a paper is to introduce a new measurement technique. Authors should expect to be challenged by reviewers if the results are not supported by adequate data and critical details.

5) Before submitting a paper, the submitting author is responsible for obtaining agreement or signatures of all co-authors and any consent required from grant sponsors or agencies. A corresponding author usually is the submitting author, who is supposed

to reply technical questions after the paper publication.

6) The length of a submitted manuscript should contain new contribution. For example, an obvious extension of previously published work might not be appropriate for publication.

7) After paper acceptance, the authors will receive an email with specific steps regarding the submission of final version. Please follow these steps to avoid any delays in publication process. Most journals require that final submissions be uploaded through online system for easy tracking. Final submissions usually include high quality graphic files, formatted word and pdf version of accepted manuscripts and contract information of all authors. Feel free to contact the administrative staff of journal for any concerns or questions.

4.1.9 Some Common Mistakes

1) The word "data" is plural, not singular.

2) The subscript for the permeability of vacuum μ_0 is zero, not a lowercase letter "o".

3) The word "alternatively" is preferred to the word "alternately" (unless you really mean something that alternates).

4) Use the word "whereas" instead of "while" (unless you are referring to simultaneous events).

5) Do not use the word "issue" as a euphemism for "problem".

6) When compositions are not specified, separate chemical symbols by en-dashes; for example, "Ni-Mn" indicates the intermetallic compound $Ni_{0.5}Mn_{0.5}$ whereas "Ni-Mn" indicates an alloy of some composition Ni_xMn_{1-x}.

7) Do not use the word "essentially" to mean "approximately" or "effectively".

8) Be aware of the different meanings of the homophones "affect" (usually a verb) and "effect" (usually a noun), "complement" and "compliment", "discreet" and "discrete", "principal" (e.g., "principal investigator") and "principle" (e.g., "principle of measurement"). Do not confuse "imply" and "infer".

9) Prefixes such as "non" "sub" "micro" "multi" and "ultra" are not independent words; they should be joined to the words they modify, usually without a hyphen.

10) The abbreviation "i.e." means "that is" and the abbreviation "e.g." means "for example" (these abbreviations are not italicized).

11) Use a zero before decimal points: "0.25" not ".25" .Use "cm^3" not "cc" .

12) Indicate sample dimensions as "0.1cm × 0.2cm" not "0.1 × 0.2cm^2" .

13) The abbreviation for "seconds" is "s" not "sec" .

14) Use "Wb/m^2" or "webers per square meter" not "webers/m^2" .

4.2　Letters for Academic Communication

With the rapid development of information technology, the application of Internet has gradually permeated all aspects of society. Bill Gates, the former president of Microsoft, believes that information technology and the Internet have become an important part of life, and that people can use the Internet to communicate with each other. E-mail, known as the mother of Internet applications, is a way of communicating in a computer medium. Nowadays, E-mail plays an irreplaceable role in life. Email, or E-mail, is the most common means of communication used in e-commerce today. E-mail is easy to use, fast and efficient in delivering information. Because E-mail has both oral and written language characteristics, it has become a new and common communication method in people's daily life. Especially in scientific research, the ability to write letters in English is the basic ability you need to do research. Good letter writing ability is an important asset, and it will become an important factor in your success in international academic exchange. In English writing, you should pay attention to the sentence structure, words, phrases, grammar, punctuation, logical thinking ability, the structure of the material, and the format of letters. The following introduces a letter format—block style.

　　This style is very popular, because it can save typing time. Since all rows in this format start from the left distance, there is no need to count or center. Write in block letter, the space between the text and the call is one or two lines. There is no space needed for the first sentence of each paragraph, but one or two lines are needed between paragraphs. The signature can have two styles. The first one is written on the lower left, which is the most common and formal. It can also be written on the lower right, which means that the relation between the writer and the addressee is more familiar and casual.

4.2.1 Contact Emails for Graduate or Post-doc Positions

In this section, you usually briefly explain your motivation and catch the attentions from your targeted professor. A clear email title with key words, such as "Graduate scholarship applicant" or "Post-doc position application" should be written. Since a professor will receive averaged 200 emails per day, if the email title is not clear, the email will not be noticed by the professor. In the email context, some important information, such as research area, current degree and affiliation, available date, financial support or current supervisor name, should be included. A detailed copy of resume/CV and 3—5 reference names should be attached in this email, so the professor may download them if he/she has interest in this applicant. If the applicant has ever been granted some honours, such as the best research paper award, best researcher nomination or best design contest awards, these awards should be listed and highlighted in the email to emphasize the research achievements. If the applicant reads through recent publications of this professor and has potential idea to collaborate in future, it is recommended to write down these future work plans to show his/her research potential. If there is no response from this professor in around two weeks, the applicant may contact the professor again. Don't keep sending emails to a professor too frequently, for this manner is annoying.

Below is a typical example of such a letter for a post-doc position application.

Title: Post-doc position application from ××× University

Example 1: Hello Dr. ××,

I am a recent Ph.D. graduated from ××× University (West Lafayette, Indiana) with the specialization in structural mechanics and energy harvesting systems. My dissertation research was focused on the vibrational energy characteristics and energy harvesting optimization of a disk-type MEMS sensor in various fluid environments. As I am currently looking for a post-graduate research position, I wanted to know if you have an open position in the field of ××× or other similar areas that may be relevant to my background. I appreciate to get financial support from your research projects. I am attaching a copy of my CV along with a list of 3 references to this email in case you would like to take a look at them.

Thanks,

×××

Example 2: Dear Professor ×××,

I am a graduate applicant for Ph.D. degree in Electrical Engineering from ××× University in USA. Fortunately, I have been admitted as a Ph.D. student for the fall 2017 into your University. I got my M.S. degree in Electrical Engineering (in Control Major) in ××× University, Greece. My master thesis was about "stability analysis, modelling and optimization of discrete time systems". I also got my bachelor degree in Applied Mathematics with a GPA of 3.8/4 and a rank of top 5. As my resume shows, I have a notable background and strong interest in Mathematics in Control and Optimization. I have a few publications in the areas of ① Comprehensive Investigation of Stability Analysis of Time-Delay 3-Order Systems, ② Model Predictive Control using Artificial Intelligence.

As I have common interests with your field of research, I wonder if you have any TA/RA opening position available for a Ph.D. student at the beginning of fall 2017 semester or spring 2018. I will benefit from knowledge and experiences of such a great professor like you. In order to supply more information about myself, I have sent you my resume in the pdf format, which entails my education history and past research experience. For any more information about me, feel free to reach me by phone ×××-×××-×××× and by email ×××@gmail.com.

I'm looking forward to hearing from you.

Best Regards

×××

See Appendix Ⅰ—Ⅲ for details.

4.2.2 Cover Letter for Faculty Position

Every year, many universities announce application openings for faculty positions. Most of academic positions are full-time, tenure-track assistant professors. Details of such job requirements include: education and experience qualifications, robust academic records, skills and abilities, research area preference. Ph.D. in the required major or closely related field is one of basic qualifications. Usually excellent oral and written communication skills are highlighted. To apply for such faculty positions, an applicant should prepare and submit a cover letter to briefly introduce himself or herself. Some evidences are supposed to be listed in this cover letter to impress the faculty search committee members. Below is one good example of cover letters for your reference.

Example:

March 16, 2017

××× University

Dear Sir/Madam,

I am writing to express my strong interest to apply for the Assistant Professor of Biomedical Engineering. I earned my Ph.D. degree in Biomedical Engineering from University of ××× in 2010 (advisor: Dr. ×××), and MS degree in Biomedical Engineering from ××× University and BS degree in Electrical and Computer Engineering from ××× University. I worked for 7 years as a System Design Engineer in ×××, 1 year as R&D Biomedical Instrument Engineer in ×××. I have been teaching biomedical experiments, tissue modeling /characterization courses in ××× University since 2015 as a tenure track assistant professor. I finished my postdoc training at University of ××× in Biomedical Engineering and published 3 top journal papers. My research interests include bio-tissue modeling, cardiovascular mechanics, medical device development, microcirculation and hypertension. Up to date, I have published more than 20 journal papers and about 30 international conference papers. I enclose my CV, published papers, teaching statement and research statement for your consideration.

My fascination in teaching and research has been fostered by years of undergraduate teaching in ××× and ×××. My teaching career involved lectures, discussions and laboratory procedures on many undergraduate and graduate courses, such as introduction to biomedical engineering, bio-tissue modeling/characterization, EEG signal collection and analysis, and Biomedical Engineering Labs. My past experiences helped me build up professionalism and appreciation expected for a teaching environment. I am looking forward to being a part of your department with its strong respect in providing high quality of education, research and service.

Thank you for your evaluation. You can reach me by phone ×××××××××× or by email ×××@gmail.com. I look forward to hearing back from you soon.

Sincerely,

Name

Recommended bibliography:

The following is recommended to learn writing books. You can read under the class.

1) Strunk W, Jr. , White E B. 1987. The Elements of Style. 3rd ed. New York: Macmillan.

2) Whitesides G M. 2004. Whitesides' group: writing a paper. Advanced Materials, 16(15): 1375-1377.

References

Camciunal G, Laromaine A, Hong E, et al. 2016. Biomineralization guided by paper templates. Scientific Reports, 6: 27693.

Day R A. 1994. How to Write and Publish a Scientific Paper. 4th ed. Phoenix: Oryx Press.

How to Write a Research Introduction. https://www.wikihow.com/Write-a-Research-Introduction.

IEEE Journal template.https://journals.ieeeauthorcenter.ieee.org/create-your-ieee-article/authoring-tools-and-templates/ieee-article-templates/.

Kallestinova E D. 2011. How to write your first research paper. Yale J. Biol. Med., 84(3): 181-190.

Nature, manuscript formatting guide. www.nature.com/nature/for-authors/formatting-guide #a5.2% E2 % 80%8B.

Sage publishing, Manuscript Submission Guidelines. https://us.sagepub.com/en-us/nam/journal/integrative-cancer-therapies#TitleKeywordsAndAbstracts.

Swales J M, Feak C B. 2004. Academic Writing for Graduate Students. 2nd ed. Ann Arbor: University of Michigan Press.

Whitesides G M. 2004. Whitesides' group: writing a paper. Advanced Materials, 16(15): 1375-1377.

Appendix

I The resistive switching mechanism of Ag/SrTiO$_3$/Pt memory cells.

II Effects of the electroforming polarity on bipolar resistive switching characteristics of SrTiO$_{3-\delta}$ films.

III Superior resistive switching memory and biological synapse properties based on a simple TiN/SiO$_2$/p-Si tunneling junction structure.